A Practical Introduction to Facilities Management

A Practical Introduction to Facilities Management

Ian C. Barker MCIOB

Whittles Publishing

Published by
Whittles Publishing,
Dunbeath,
Caithness KW6 6EG,
Scotland, UK

www.whittlespublishing.com

ISBN 978-184995-096-1

Typeset by Exeter Premedia, India

Printed and bound in Great Britain by Bell & Bain Ltd, Glasgow

Contents

List of theory and information slots

Chapter 1

Chapter 3

Chapter 4

Chapter 5

Chapter 7

Chapter 8

Abbreviations used

ACMs	Asbestos-containing materials
AMR	Automated meter reading
AOC	Association of Colleges
BIFM	British Institute of Facilities Managers
BMS	Building Management System
CAD	Computer-aided design
CCL	Climate Change Levy
CCTV	Closed-circuit television
CIOB	Chartered Institute of Building
COSHH	Control of Substances Hazardous to Health
CPD	Continuous professional development
CPM	Critical Path Method
CRB	Criminal Records Bureau
CTR	Collective team responsibility
ELI	Employers liability insurance
EMS	Environmental Management System
FM	Facilities Management
M & E	Mechanical and Electrical
MPAN	Meter point reference number
O & M	Operations and Maintenance
PCV	Passenger carrying vehicle
PDA	Personal digital assistant (palmtop computer)
PFI	Private Finance Initiative
PLI	Public liability insurance
PPM	Planned preventative maintenance
PQQ	Pre-qualification questionnaire
QA	Quality Assurance

QDP	A national questionnaire on the quality of student perception
QMS	Quality Management System
SAR	Service Area Reports
SLA	Service Level Agreements
TRV	Thermostatic radiator valve
TUPE	Transfer of Undertakings and Protection of Employment
VRF	Variable refrigerant flow
WEE	Waste Electrical Equipment

To aid you through this book, various key pieces of information will be highlighted in the following way.

Note to self

These are action points to draw your action to key pieces of text or remind you to do something.

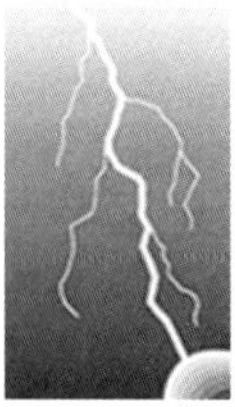

Warning
These are warnings of key things to avoid.

★★★★★★★★★★

These are bright ideas to get you thinking about quick wins.

★★★★★★★★★★

Theory and information slot

Most of the work described in this book is based on practical experience tied closely to accepted theories and information about facilities management. These slots will detail some of the theories behind my work. Theory and practice may not always be identical, but you have the option to compare the two side by side. These slots will also contain key pieces of information. The information given in relation to the theories will be summaries, not the full theory.

The information slots will work in two ways. For the experienced Facilities Manager, they will remind him or her of some of the key theories learned as a student. For the student, they will point the way to exploring the theories in greater depth.

Aide Memoire

The Aide Memoire section at the end of each chapter will summarise the key points covered.

Chapter 1
Getting started

What is Facilities Management?

Facilities Management (FM) can be defined in one simple phrase: "everything that is not core business." Practising Facilities Managers must be prepared to turn their hands to an increasingly wide range of areas, some specialist, to allow the core functions of a company to function effectively.

This "everything" can differ from business to business and new areas are being added to the facilities mantle every day. Facilities Management services are often seen as the natural home for everything that is not the core activity, although, in general, data and management information systems, information technology, human resources and finance are defined as separate services.

For me, one of the attractions of going into FM as a profession was the variety that it provides in terms of job content. You could be doing a different job every day of the week, so there is never an opportunity to become bored or stale. Facilities Management is a developing area and requires the practitioner to be on top of their game at all times. You will be looking at new things on a daily basis and you will probably be expected to achieve a high level of knowledge quickly and be able to adapt that knowledge to suit the ever-shifting sands of FM.

As an FM practitioner, it is essential that you are highly organised and have robust systems in place to meet this requirement of remaining up to date and maintaining your Continuing Professional Development (CPD). This book aims to guide you through setting up robust systems that overlap all the FM functions and ensure that there is no over-duplication of your work.

As you work through this book, you will be introduced to some of the theories of FM through the use of examples. I have taken this approach because, as you become a practitioner of FM, you will quickly realise that there is no "one size fits all" answer to problems. There are tried and tested theories

Theory and information slot: the future of FM

Facilities Management has come of age and is here to stay. Increasingly, Facilities Managers are being viewed as strategic managers of the organisation they represent. They have always been seen as key members of the organisation's team, but they are now being seen as having the ability to affect the bottom line of the organisation, rather than as a resource-sapping function.

The strategic stewardship of an organisation's assets during times of austerity in particular brings confidence to an organisation, allowing them to focus on core activities rather than worry about the peripheral FM functions. It is this stewardship that can reduce risk to the organisation while adding positive value to the organisation's estate.

Increasingly, the role of sustainability manager is sitting within the facilities function and, again, in times of austerity, social responsibility in the form of proactive sustainability can provide an added competitive edge to an organisation.

Note to self

Remember you are in a developing industry. However, the sheer diversity of functions allows innovation to drive the FM process forward.

on FM, but rarely do all the pieces of the jigsaw fit together exactly. You will find that you may need to adapt, modify and fundamentally change accepted practices to make them work effectively in your own organisation. This is one of the fundamental threads that run through performance management and the Plan–Do–Review–Revise management mantra.

I cannot emphasise enough that it is essential that an FM practitioner is trained. There is no substitute for understanding the principles of both general management and specialist FM and using those theories and practices alongside this book to manage your portfolio. A fully trained FM professional is an essential prerequisite in today's fast-paced FM environment.

The value of experience

Many older FM practitioners gained their professional knowledge through the "school of hard knocks". Making mistakes is part of any learning process.

However, today's new recruits to FM do not have the luxury of being able to make mistakes, as today's pressurised environment means any mistake could have a detrimental effect on the core business. A key example of this occurred in 2012, when a series of IT crashes in the banking sector locked customers out of their bank accounts for days. Ultimately, this will cost the banks millions of pounds. This example shows how businesses can sometimes be dramatically affected by the loss of a separate support service, although, in this case, not FM.

You must understand, however, that you will inevitably make mistakes. A good FM practitioner will use this to his or her best advantage and come out of the experience far richer. The FM practitioner who says "I don't make mistakes" is probably lying. All you can really do is learn from the mistake and make sure it does not happen again.

This book has been developed as a helping hand in an area of management where it is all too easy to become overwhelmed by the sheer size and volume of the role. Experience is the one thing that cannot be taught. Your experience will grow on a daily basis. Never be afraid to try new things and to make mistakes, as this will make you a stronger FM practitioner. Equally important, however, is never to be rash with your establishment's assets. No one appreciates a gung-ho attitude, and certainly not your staff, who will be looking to you to create stability, not chaos.

When you move from one company to the next, you take with you the best work practices you have seen and developed, and leave behind those practices

Theory and information slot: hard and soft services

When jobs for Facilities Managers are advertised, they are often grouped under the titles of Facilities Manager (Hard Services) or Facilities Manager (Soft Services), or an amalgam of the two (Hard and Soft Services). But what are "Hard" and "Soft" services?

Hard services generally relate to all mechanical and electrical services within a building. Sometimes referred to as M & E, they cover the specialist functions of heating, ventilation and their control systems. The Hard Services Manager invariably comes from a heating engineering background, or has been specifically trained to deal with hard services.

Soft services relates to most of the other trades that run through FM: cleaning, porterage, post room and maintenance disciplines such as painting and decorating or joinery. A Soft Services Manager is more focused towards personnel management than the Hard Services Manager.

and policies which did not work. Experience is therefore a key element in your FM armoury and you should aim to build on your experience in a logical way when considering your CPD. Work closely with your professional institution to ensure that not only do you develop as a truly outstanding practising Facilities Manager, but that you also learn from your peers within the organisation and pass on your skills through networking. Choose your jobs wisely to encompass as wide a range of activities as possible. You will find that, as your skill levels improve, you will be able to apply your experience in a number of ways and use it to cross over into other disciplines, thus broadening your portfolio. Remember that there is no substitute for tried and tested FM and general management training.

The trainee

We don't all start in FM as managers. The beauty of FM is that there are so many different levels that the trainee can start work in one of many disciplines and steadily work their way up while gaining that all-important key experience. The Trainee should have their eyes fixed on whatever level they ultimately see as their ideal position and use the skills of the personnel around them to enhance their own skills. Building that key experience is very important if you aspire to be a good all-round FM practitioner. Be sure that, while building your key experience, you also develop your academic side. Again, FM is so diverse that you can pick up FM skills in modules or batches of learning. I cannot emphasis enough that, to become a well-rounded FM practitioner, you need to understand FM thoroughly and you can only do that if you understand and have had a thorough grounding in theory. Very much like the quote "a dog is not just for Christmas," FM is for life! Once it is under your skin, you cannot think of an alternative career.

Warning
If you take on too much at once, you are likely to fail. Do things in bite-sized chunks.

Do you know that bloated feeling? When you go out for a celebratory meal and are faced with a banquet of new and interesting things, you have a tendency to over-eat, leaving yourself with that bloated feeling. When you become a Facilities Manager for the first time, you will suffer from that bloated feeling for a considerable period of time. The sheer volume of information, varieties of work, compliance constraints and pressures and, in particular, the diverse set of staff that you require, will at times push you to the edge of despair. You need to be able to control your workload effectively and not get that bloated feeling that will weigh you down, and wear you down, unless you can master and control it. You will ultimately be your own judge of whether you are successful. Don't be too harsh or too demanding on yourself in those formative months, or indeed years, and allow yourself time to review your progress every step of the way.

Let's start by setting a framework to follow. Don't set off like a firework, shooting in all directions. You must be measured and practical about every area that you tackle.

Aims and objectives: setting out your stall

Don't set off by ploughing your way through your workload without taking a step back and evaluating what it is that you are actually doing. Never assume that your predecessor had all the pieces of the jigsaw. Your strategy should always follow the same format: identify, evaluate, plan, action and review.

Identify

You first need to sit back and look at the services you provide.

- *What.* What services do you provide and are there any other services that you should be providing?
- *How.* How do you provide the services that you are responsible for?
- *When.* Obtain an overview of the times that you provide your services. You will probably not be there for the full operational day. Make sure you know what is going on when you are not on site, and make sure that these do actually take place. You will often find that services that should be provided at the end of the working day simply do not happen and that actual working hours do not match contractual hours. Remember that the provision of FM services may be 365 days of the year, 24 hours a day. Give equal merit to all aspects and parts of the working day.
- *Why.* Find out why you are providing particular services and what the rationales behind the services are, e.g. staffing levels.
- *Where.* Understand where services are provided from and why these locations are used. Are they located in the best place for ease of access and delivery?

Evaluate

The next stage is to evaluate whether all these services are essential and to determine any gaps within the service. It is clear that as a new and forward-thinking manager you will want to develop your service to fit the requirements of your organisation. When you have got to grips with the service you have already, you can begin to understand how best to provide it. To see where the gaps in the service are, you need to have a full overview. You also need to know the aims and aspirations of the organisation in terms of where they want your provision of services to go.

Take time to understand the connections between services and to check that there are no unnecessary overlaps. There may be instant gains to be made by putting two services together to create stability and continuity of staff during illness and holidays. For example, putting the roles of Helpdesk and Post together can provide development opportunities for staff. Allowing both teams to increase their skills to provide stability during holidays also allows your staff to develop more interest in their roles and increases team spirit.

What is evaluation?

"Evaluation is a careful and measured appreciation of the subject area, creating an understanding of the subject as a whole."

Evaluation is a measurement tool and sometimes it is difficult to assess something if raw data are not available on which to base decisions. Sometimes you can be "flying in the dark" or basing your decisions on gut instinct or staff perceptions on how good or bad a service is.

Your evaluation of the service should be an ongoing process. What you are aiming to do is draw a line in the sand and be able to quantify the success of the department in its entirety. You don't want to have to replicate this exercise each year, so set up evaluation systems into which you can feed data to give you an appreciation of how far the service has successfully travelled. This needs to be one of your first tasks: effectively drawing a line in the sand by bench-marking your service. Defining the service in this way will give you a firm base from which to measure your success. Later in the book we will look at Quality Systems and these will help you to pull together evaluative tools that will stand you in good stead for the future.

What should be included in the evaluation of services?

Do not become too insular about evaluating your service. It is sometimes too easy to completely miss two very important parts of the jigsaw: the service that you provide and the people you are providing the service to.

Do not forget that you are one of the biggest cogs in the service you provide. Remember that if your service fails, it could be your fault. This is not always the case, as external influences inevitably contribute to failure;

however, you are the person in the driving seat and if the car crashes you could be held accountable.

Ensure that your processes are also evaluated and, more importantly, formulated to fit the service. Ensure that you leave an audit trail and document your evaluation in a coherent way. Sometimes you may get to the end of an evaluation and be challenged as to how you arrived at your conclusions. If you have not set your work out in a way that can be audited, you may struggle to justify your decisions if you cannot refer back to a set of coherent notes. You have the power to shape your own department, but, if your systems and processes for protecting the department are not robust, then the whole system is likely to fail.

Without the people you are providing the service to, there is no service to provide. A large proportion of your evaluation must include these people. Take time to speak to your clients and customers and make sure that their expectations of the service are being met. You may find that your service is wide off the mark in terms of meeting the customer's expectations. Engage with staff and visitors. If you are in an education environment, speak to both staff and student groups and make them feel from the outset that your job is to provide a good service for them.

You may be lucky and drop into a service that is fully defined, e.g. a Private Finance Initiative (PFI), and this will certainly make your life easier. However, it is more likely that there will be no specification defining your service capabilities and requirements.

Your organisation's strategic objectives will provide you with an invaluable guide to the service you need to provide. Do not, however, become drawn into promising too much in terms of service early in your tenure, before you can evaluate whether it is practical to do so. Under-promise and over-perform is a useful guide.

Sources of information for your evaluation

There may already be good, valuable sources of information for you to tap into as part of your evaluative process. These may include some of the following sources.

- *Internal departmental reports*. You may find that your predecessor has already evaluated certain aspects of your service. I am always suspicious of previous reports, not because I believe they are inaccurate, but because they rarely look at the whole service and can be one-sided, promoting a rosy picture. This could be misleading and make your job more difficult in the long term.

- *Questionnaire Data Processing (QPD).* This is an independent company providing statistical data to educational establishments. It is a national questionnaire undertaken in colleges and compares certain aspects of college life between establishments. It can be very useful in terms of identifying areas such as cleaning, hygiene and security, which can influence QDP scores.
- *Internal quality reports.* Most quality systems will produce internal reports which may identify issues and suggest solutions.
- *Service Assessment Reports (SARs).* Again, in a college environment, all services including the core curriculum are judged on the SAR process. The SAR process defines the targets set for each aspect of service provision and periodically tests progress against those targets through progress meetings each term. This effectively becomes a self-assessment of the service provision and strategic objectives of the organisation.
- *Customer feedback reports.* Key to any organisational success is the way in which the organisation treats its customers. These reports can be invaluable in understanding what customers feel about the service.
- *Audit reports.* Unbiased opinions from external auditors can help you to see where improvements need to be made.

Evaluative conclusion

It is important that, once you have evaluated your service provision, you reach a conclusion on the effectiveness of the service you are providing. This is the prime reason you began the evaluation in the first place. Take time to reach a summarised conclusion which identifies the key strengths and weaknesses of the department. Use of a "ragged" report (red, amber, green) will enable you to achieve the following:

- see where your strengths are and how importantly you rate those strengths in relation to the overall service provision
- identify your weaknesses and see where those key weakness affect the service in a detrimental way
- prioritise your actions
- set time-scales

Table 1.1 Example of an evaluation of services report

Area	Strengths	Weaknesses	Area for improvement	Date for completion	Progress at Aug 2012	Progress at Jan 2013
Helpdesk	Fully functioning section with well-defined software Motivated staff	Software not being used to full extent Helpdesk not in high profile area	Training required to ensure software being used fully Move Helpdesk to front of house	Jan 2013	A	
Condition survey	50% of stock rebuilt within last three years	No full condition survey undertaken	Undertake full condition survey	Oct 2012	R	
Health & Safety	Full risk assessments undertaken Trained staff Successful audit undertaken with positive outcome	Risk assessments need reviewing in Oct this year	Prepare for and calendar review of risk assessments later in year	Oct 2012	G	
Utilities	Automated meter reading (AMR) in place Targets set for reduction	No-one monitoring AMR Utilities contracts up for renewal Targets not being monitored	Delegate responsibility for reading AMR Prepare tender for utilities Make the person responsible for the AMR also responsible for reporting	June 2012	A	

R = red, A = amber, G = green.

Plan

Now that you have successfully evaluated your systems and service, you should be in a position to have a general overview of the service that you provide. You can't just flick a switch and see everything fall into place. Installing a coherent system within a busy FM department will depend on the size of the institution.

Look to install and set up your basic system within the first year and then allow a further two years to evaluate and review your progress. This may sound like a long time, but the time will fly by quickly. You will obviously want to make a mark quickly in some areas, but keep a long-term strategy

Table 1.2 Example of a time-scale of plans for improvement

Area	**Area for improvement**	**Apr**	**May**	**Jun**	**July**	**Aug**	**Sep**	**Oct**	**Nov**	**Dec**	**Jan**	**Feb**	**Mar**
Helpdesk	Training required to ensure software being used fully												
Helpdesk	Move Helpdesk to front of house position												
Condition survey	Complete condition survey												
Health and Safety	Review risk assessments												
Utilities	Secure responsible person												
	Prep tender for utilities												
	Set up reporting structure												

key to your development plan. Always remember that you need to keep an eye on the future horizon, so having a long-term strategy is essential.

Look initially at the things you can implement quickly and perhaps start with those, but always have a plan. Don't just aimlessly try to install processes without setting yourself objectives and a time-scale, otherwise you will never achieve results.

Immediately you can begin to see that you have made room in your diary for the work needed and you have some clear deadlines to meet. Don't be afraid to put detail into your plans. The more detailed your plans, the greater the likelihood that you have covered all the basics.

Planning, Organising, Commanding, Coordinating and Controlling form part of most modern management theories and form the bedrock of moving a service forward in a productive and measurable way. You, in turn, will develop

Theory and information slot

Henry Fayol (1841–1925) considered that management consisted of six functions:

- Forecasting: predicting what will happen in the future
- Planning: developing a course of action and setting time-scales
- Organising: mobilising the work and allocating tasks
- Commanding: directing and motivating staff
- Co-ordinating: making sure all services pull together as one
- Controlling: monitoring progress against targets

Adapted from Fayol's six managerial activities table 16.2 in Buchannan, D.A. & Huczynski, A.A. *Organizational Behaviour* (7th edition). Pearson Education Ltd: Essex, 2010.

Theory and information slot: Gantt charts and the Critical Path Method (CPM) in Gantt Buchannan, D.A. & Huczynski, A.A. *Organizational Behaviour* (7th edition). Pearson Education Ltd: Essex, 2010.

Gantt charts were developed by Henry L. Gantt (1861–1919) as a project management tool.

The CPM was developed by Morgan R. Walker in the 1950s.

Adapted from http://en.wikipedia.org/wiki/PDCA

Both have been developed into other systems, but remain the mainstay of modern project management plotting.

your own interpretation of these classical subject headings, but never forget their origins and always start at the beginning.

I have already mentioned that you need to reflect the strategic objectives of the establishment and these are where you start. Lay out your policies and procedures and make sure that they reflect your strategic objectives. Policies and procedures are considered later, so I will not dwell on them now; however, they do need to be conveyed to your staff so that they understand what you are trying to achieve. Your plan also needs to be conveyed to your staff so that they understand your time-scales and, if necessary, this can be referred to if time-scales are not met.

If you do not define policies and procedures in a logical way and lay them out in an easily understandable format, your staff will, either by inaction or subversively, wreck your plans before you begin. Don't be disheartened if you fail at the first attempt. Sometimes you need a failure to help focus staff by letting them know you are determined to follow your plans through. However, never believe for an instant that you have all the answers. As with most things in life, there is an element of compromise. If you feel you are going to have to go down this route, build excess into your plans to allow yourself a compromise position which suits you, not your staff.

Action

You now need to implement your plans. Your plans should be dynamic and ready to go before you start to implement them. Do not allow yourself to be rushed in this process, because if you fail to prepare, you must be prepared to fail. Your plans can be road tested by staff before implementation and this will also create buy-in to the plans by your staff.

Buy-in is essential for any system to work. Most importantly, you should believe in your plans. Do not approach things in a half-hearted fashion because, again, if you fail to convince your staff that you believe you are doing the right thing, there is no way that they will believe it for you.

You will never push everything through at the same time. Create a strategic plan or critical path which you can follow in a step format. Undertaking this in bite-sized chunks will make the whole plan easier to follow and will not overwhelm you. Create those chunks of activity with defined starts, middles and ends. Approaching this in a logical way will ease you through the process and allow you to see defined progress. This way you will avoid that bloated feeling.

Review

Review, review, review and review again. The work you do requires constant appraisal and you will never be finished. Review is the strongest tool you can ever have and, if you link this in with your Quality Management System (QMS), it is a double winner as audit and review are good bed fellows: one identifies problems and the other solves them. Never undersell. Review and make sure it becomes part of your staff's culture to constantly reappraise what you are trying to achieve. We will look more closely at a QMS later in the book.

Note to self

Never bite off more than you can chew at any one time. All you will do is overwhelm yourself and do each of the jobs poorly.

Review needs to be an open task. It needs to be in the public domain or, rather, within the domain of the department. Everyone needs to see that the systems are working and, when they don't work, people need to see that you are not blinkered, but are prepared to change to achieve the right result. This will encourage your staff to buy-in to the process and makes them more likely to help you to amend poor performance and continually strive for a better way of working.

Some staff will inevitably kick against all that you do. After all, they may have spent years building this cosy nest to hide in. They may never have been challenged. Try to move forward positively with change: there is no such thing as custom and practice. Try not to accept the old adage "this is the way we have always done it." Inevitably, staff will point to this when they feel under threat.

Consider all the angles, but never assume you are always right. Check, review, revise. Never, ever, forget to provide feedback to staff on progress. Lack of feedback is an incentive for them to fail if they think you are not monitoring the processes you are asking them to implement. By far the best way to deal with an issue that your staff do not want to deal with, or that they protest about, is to try and make your ideas into their ideas. Try to sow the idea with them and let them come around to it by making it their idea. This way, they buy-in to the process.

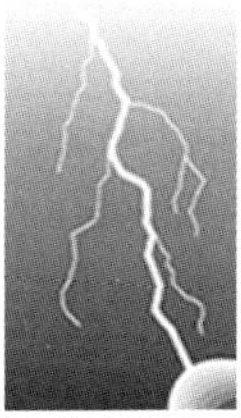

★★★★★★★★

Plant the seed of an idea you want your staff to champion and let them nurture it to fruition as their own idea.

★★★★★★★★

Getting started

How to get started? Let's remind ourselves of the process needed. You will see this process of Identify, Evaluate, Plan, Action, Review throughout the book and it will assist and guide you as we go along. This process will stand you in good stead as we assemble your FM function. It must be stressed that I am not trying to imply that all FM processes are flawed and that you should deconstruct them just for the sake of it, but, as with everything in modern life, if you stand still, then you fail. Continuous evaluation and review of your services can only be a good thing. Even when you feel you have got it right, there is always room for improvement. Standing still is not really an option in today's fast-paced society. You need to keep striving for improvement in the way you provide your services.

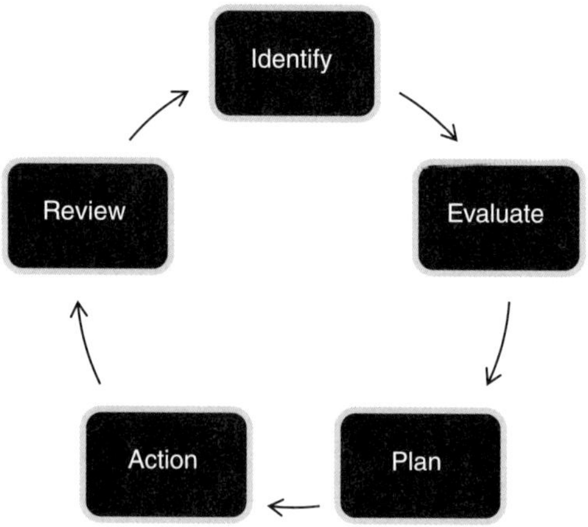

Try to bench-mark your service as you go along. This is by far the best tool you have, as, if bench-marked accurately, you will be able to plot improvements in service and convey those improvements both to your staff and to the wider organisation. If you do not feel comfortable with this methodology, adapt it to fit your work environment, or simply use the tried and trusted Plan, Do, Study, Act (PDSA) methodology.

Theory and information slot

Plan, Do, Study (check) and Act, or PDSA, was introduced by the renowned statistician Edwards Deming. The PDSA cycle is a continuous quality improvement consisting of four repeated steps for continuous learning and improvement. It is sometimes known as Deming's Wheel, or the Continuous Improvement Spiral.

Adapted from http://en.wikipedia.org/wiki/PDCA

Chapter 2
The basics

Tools of the trade

No worker can work without tools. Set out your stall and ensure you have the tools of the trade.

Computer systems and software

You will obviously need a PC. The PC must be capable of running a number of applications, so make sure you get the highest specification of computer possible. Speak to your IT department and tell them what you want from the system and what programs you are likely to run. You may find that the number of programs you wish to run on the system will slow it down, leading to frustration. Planning ahead will allow you full functionality.

Current FM almost always demands the use of technology and you will find that you need to run a wide variety of software on your PC, from Computer Aided Design (CAD) programs, Helpdesk and financial software, to practical software solutions such as temperature monitoring, a Building Management System (BMS) and environmental systems. Make sure your PC is matched to your requirements.

Try to avoid buying a printer in the first instance. If you set yourself up in the correct way you can usually avoid the need for printing. This will help with your sustainability obligations. Start with systems that follow good practice and you will naturally follow this philosophy and the "need" to print documents will reduce. Information is king, but it can also paralyse you.

Phones

One of the best resources I have is my phone. Yes, it seems obvious to have a phone, but today's phones have more than one function. Modern phones are more like Swiss Army knives! They can be used for:

- making phone calls
- linking to your calendar to provide alerts for meetings
- linking to your emails to allow quick responses to queries
- Internet surfing, again allowing quick access to information
- access to your home area where data or information is held
- making voice memos on site that can be listened to later
- clock or timer
- various Apps, such as levels, compass, lights, taking photos

Administration structure

Information, information and information. If you don't get this flow of information right from the start, you will be swimming upstream forever more. Most of the time the FM practitioner will be coming into an environment that has already had an FM function for a number of years. You may be faced with mountains of paperwork and archived material that is neither accessible nor in a logical form. Your first task is to sort this out and start building your FM strategy.

Note to self

Use technology to the best effect. Think about what you want in advance rather than trying to fit it all together afterwards.

Note to self

Pull together all the information, lay it out in a logical format and fill in the blanks.

You cannot do everything at once: you are not superhuman, although you may think to the contrary. Do not try to adapt the systems you are inheriting. Start from scratch and then, over a period of time, try to integrate your information into the new system. However, you may find that parts of your inherited system work quite well. Don't shut your eyes to this. Part of being a good FM practitioner is having the ability to recognise when something is working well and leave it alone. As you gain experience, you will find that the best way forward for any FM practitioner is to start with something that best fits your needs, then steal ideas from other FM practitioners. Do not reinvent the wheel each time you approach a problem. Develop a networking philosophy and let your peers help you. This book aims to give you a framework on which to base your best practice. Never forget the value of management theory and make sure you utilise academic expertise to the full and use the principles and theories you have learned.

If you get your administration structure right, you will save yourself lots of time and effort. It is all about getting the maximum results from the minimum effort. You already have too much to do without making it harder by overcomplicating your admin support. Concentrate on the key aspects of your set-up and plan the way your systems are likely to run in advance. Avoid duplication whenever possible, although in practice this is rarely completely achievable.

Give the layout of information on your PC some real thought from day one. The way you access and save your work over a period of time can leave you with a chaotic mishmash of documents that don't form a logical format. It then becomes increasingly difficult to locate the files you want and to access saved work.

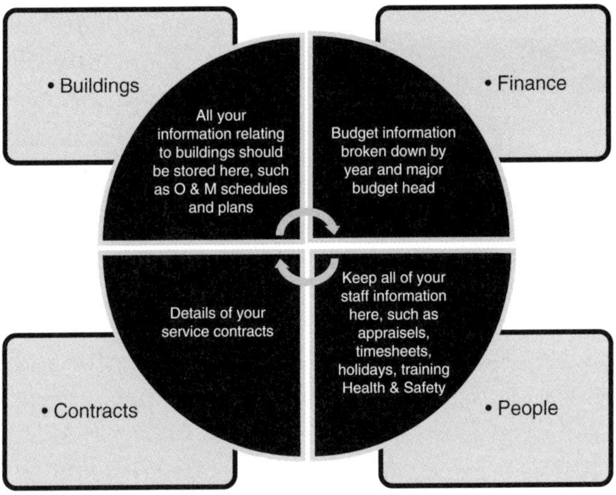

Start by breaking things down into streams of information as illustrated, but make sure that your process has seamless integration between each of the subject headings. Only by laying all of your workload out in a logical way will you ensure that you don't miss things. Just as importantly, you will ultimately find it easier to find things saved onto your PC.

Networking

Professional FM practitioners are rarely precious about their ideas and will generally give you ideas for systems, procedures and policies without obligation. As mentioned earlier, steal your ideas from people who have tried and tested systems and found them to work. If you become obsessed with this being all your own work, you will become so bogged down that you will not be able to move forward in a positive way. Networking can be the key to unlocking potential ideas, creaming the best ideas, or tweaking those you have already to gain the market edge.

Every area of the country will have an active FM group who meet on a regular basis to discuss current issues and to share ideas. Tap into this by contacting groups such as the Association of Colleges (AOC), the Chartered Institute of Building (CIOB) or the British Institute of Facilities Management (BIFM). All these organisations have special interest groups dealing with the specifics of areas you may wish to look at more closely. If you are paying the subscriptions to these institutions, then use them. Treat these FM practitioners as your resource library. After all, they probably did exactly the same when they started their new roles.

Be forceful. FM professionals are your best resource and will invariably part with their most prized possessions: systems and policies and procedures. Simply Google all the similar institutions in your area and ring them up. Remember, however, that as FM professionals are not precious about their systems, you must be prepared to part with your equivalent prized possessions. Not only does this cross-fertilisation help you to establish new and better systems, it can also help to forge friendships with your peers that can last a long time and be mutually beneficial.

> **Note to self**
>
> Don't be too proud to pinch ideas from other people.

Going out and looking at other institutions can be rather off-putting. You may come back with a sense of foreboding about the task ahead and sometimes think you may have taken on too much. This is good. It means you are starting to become self-critical about what you do. It means you are beginning to see that things need to change and that all in the garden is not always rosy. Only through being self-critical can you improve quality and take positive steps towards change.

Don't forget to use technology. There are many social networks such as LinkedIn that have special interest groups or forums where open discussions on FM topics can take place. I would be lost without the resource of the Internet. You can find almost anything you wish on the Internet if you use the correct terminology. We are increasingly using technology to reduce our reliance on paper-based systems. Using tablets and iPads to store data and minutes or notes of meetings helps us to achieve this. Syncing will allow you to integrate your iPad into your home area on your PC and allow you access to internal networks.

Being a "newbie"

As the newbie, you will get away with a lot because you are new. You are mostly also given the latitude to change things and usually an open ear from your senior managers. This is your opportunity to set out your stall and make an impression. Take advantage of this: be as forceful as you can and push things through while you have the opportunity. Once you have been there a while you will lose the ability to be radical as people get the measure of you.

Having a clear out

You will have inherited a lot and your predecessor will have left his or her mark on the organisation. This will be reflected in everything that you do and will come back to haunt you for years to come unless you are decisive. The following areas should be looked at as soon as possible.

- *Letterheads.* Is your predecessor's name on any of the letterheads? This is the first thing to get rid of, even if it means you don't have your stationery for some time. Don't send your name out on your predecessor's letterhead. It can instantly give the impression that you are not strong.
- *Emails and telephone answering machines.* Get rid of them as quickly as possible. I have fallen foul of this, having had the previous Facilities Manager's voice on the emergency phone for over two years. No one wanted to tell me it was still on there. You need to find out for yourself.
- *Internal documents.* Documents such as policies and procedures may all carry the digital signature of your predecessor. Make sure these are updated. If you do this under the guise of review, then you can then legitimately put your name to these documents.
- *Signage.* You may find your name needs updating on signage around the buildings. This is by far the best way of getting your name embedded into your organisation's culture.
- *Organisation staff.* Don't quietly creep in through the back door. Make sure you introduce yourself to each member of the management

team. These are the people you will be working for and a bit of early customer feedback can often set you on the right track.

- *Contractors.* All your contractors need to know who is now steering the ship. Make a start by inviting them all in to meet you. Forming early bonds with these people could prove beneficial in the future.
- *Suppliers.* These are just as important to your team as the contractors who work for you, so let them know you are in charge.
- *Partners and local organisations.* You may have affiliations with groups, subgroups and local authority partnerships. Making it clear there is someone new in charge can get you invited along to important networking activities.
- *Police and emergency services.* A good working relationship with the police is important. Developing an early relationship with the police helps you to tap into initiatives that may be run from time to time. This is particularly important in a town centre organisation where crime is likely to be on your doorstep. These early relationships with the police may get you a more sympathetic ear if you do get issues on your site.

The quick fix

Day one of your new position usually means stepping into your new office with a sense of anticipation and dropping yourself into your new chair at your new desk to begin to formulate your ideas. Here are a few quick fixes to help you along.

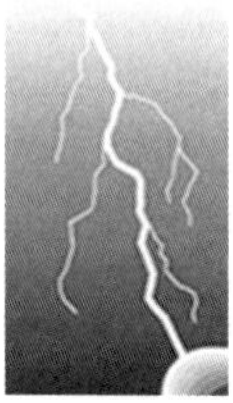

★★★★★★★★★★

It is worth the time and effort to obtain a full set of building plans. These will pay for themselves over and over again.

★★★★★★★★★★

- *See your staff.* Get this important task out of the way as soon as possible. You need to start to build relationships with these people as soon as you can. They are going to be the ones who fulfil your ambitions for the organisation. You need them to know that you are not a faceless boss who sits behind his or her desk all day. Book regular time in your diary during your first few months to get out of your office and see each of your staff in turn. It is much better to see them in their own environment. Let them know that you are there to help them and tell them how they can get hold of you when they need to.

- *Move some furniture.* What better way to exert your influence than to change your office around? Change the position of the desk or filing cabinets, or even get the room decorated. This gives out a message to everyone that this is a new regime, a new era, and that changes are going to be made, as well as stamping your authority in your space.
- *Get some plants.* This not only takes the edge off a utilitarian space, but can have a calming effect on yourself when you become stressed.
- *Get yourself a coffee machine.* Not only will this provide you with fresh coffee, but it makes the office smell friendly.

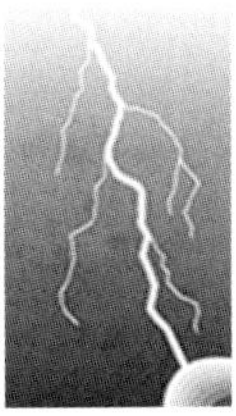

★★★★★★★★★★

Get the new room painted, put up some pictures or your qualification certificates.

★★★★★★★★★★

- *Get rid of clutter.* Remove any clutter and artefacts from your predecessor. Try to create as much space as you can in the room.
- *Contact suppliers and contractors.* Ring around all the contractors and suppliers that you currently use and start a dialogue with them. You will probably find this a useful exercise as they will want continuity of service and will want to impress you in your early days. Ask the blunt question: how can I reduce the cost of the service and increase the quality? It's a bit of a conundrum, but it will immediately let the contractor and supplier know you mean business. You may be surprised to find that they will come back with the some good ideas on how to save money or improve the service, leading to a quick win for you. However, be aware that large complex organisations are sometimes seen by contractors and suppliers as easy targets where they can cream maximum profits from minimum service. I am perhaps not being truly fair to all contractors and suppliers with that statement, but in my experience we often get taken for a ride in terms of value for money. Be cautious.

Understanding your portfolio

You need to fully understand the requirements of your portfolio. This may well be one of the most challenging things you will do. Sometimes it is difficult to even know where to start. You will probably never know everything there is to know about your site. It is essential that you adopt a strategy in approaching this initial information-gathering exercise. Begin with your physical assets.

To be able to know what you do, you need to know what you have. Sounds strangely too easy really, but unless you start out on the right foot, you will be dogged by duplication.

Physical assets

In the first phase of information gathering, this refers to the building stock and, in particular, the fabric of the building, not mechanical and electrical (M & E) assets.

Plans of your building stock are probably your most useful resource when planning any aspect of work. These plans should be in CAD form or another suitable format. Many aspects of your work will rely on accurate, up-to-date plans. If you are unable to draw up the plans yourself, commission someone to undertake this process as it will pay dividends in the future. An up-to-date set of plans will allow you to do far more than just look at layouts and will save you money over and over again, in some cases paying for themselves many times over. I have found my plans to be the best tool I have ever had.

Your plans will allow you to do the following.

- See what space you have and how that space is set up.
- Work out the gross internal area (GIA) and notional internal area (NIA) of your building stock. This is particularly useful in a college environment for your Emandate returns (Emandate is a national comparator for estates costs used by colleges throughout the UK; it allows comparison against your peers for all core FM costs, such as reactive maintenance per square metre).
- See where particular departments are and determine their size and layout to make comparative analysis.
- Create layers for M & E layouts and to build up a portfolio of working drawings.
- Define contractual borderlines for services such as cleaning, catering and security.
- Manipulate drawings to create scenarios, e.g. what would happen if I knock this wall down?
- Create defined locations for your software solution helpdesk.

Please remember, if you have spent time and money getting your drawings up to date and easily accessible from your PC, to make sure you keep them up to date. Make sure you do not duplicate or set up alternative versions of the same document, or all this effort will be compromised. Ultimately, this would mean that you are never really sure which drawing is the live one and which is the alternative version. Keep all your drawings in one place on your PC and, if you do need alternative versions, make sure they are in a completely different area on your PC.

Layers

What is a layer? A layer is exactly what it says: it is a level in a drawing. For those of you who remember when all drawings were on paper, you would provide an overlay on tracing paper which, when placed over the drawing, allowed you to see, e.g. a pipe layout. This was to avoid cluttering drawings with too much information. In CAD, a layer does exactly the same thing. It allows you to display information which would otherwise clutter the main drawing, making it too difficult to read.

Layers can be created in CAD for different levels of the drawing, e.g. M & E layouts. These can be turned on or off depending on what the viewer wants to see. This principle can be extended to all the services provided on site. With an up-to-date set of plans, layers can be added to indicate the following information:

- fire plans and escape routes
- cleaning schedules
- contractor information, such as access and egress points, parking places, waste drop off and pick points, specialist kit layouts and positions
- planned preventative maintenance (PPM)
- measurements and dimensions

Once you have understood the physical elements of your building stock, you need to look at the internal fixtures and fittings of the building. These can be layered over your plans if required.

Internal fixtures and fittings can be items such as furniture, either fixed or movable, or M & E plant. Never underestimate the value of putting all of your M & E kit onto your layer plans. You will find so many uses for this information, e.g. when putting together tenders for maintenance or replacement, to include in guidance notes to staff, or simply to give yourself an overview of where your plant and equipment is located in your plant rooms.

Operation and maintenance manuals and plans

Operation and maintenancc (O & M) manuals are your diary of the individual building. They are also a way of keeping all the information relating to a building in the same easily accessible format and in a style that can be worked through in a logical way.

Typically, an O & M manual will contain:

- details of the build, such as who built it and the name of the architect
- statutory details, such as permissions and building control certificates
- details of all the assets in the building, such as pumps and boilers
- details of manufacturers, manuals and warranty details
- PPM details for the entire building
- certifications, such as fixed wiring tests

A typical O & M manual can be expanded to deal with items such as service contract details and historical records, and details of checks and inspections.

Your building stock will generally contain a mixture of buildings of different ages and in different conditions. Your O & M manual should be organised by building and set up in a uniform way so that no matter what building you are looking at, you will find the information in the same place.

To help in locating information, it is better to keep your building information in building-specific files rather than, for example, putting all the plans in one file and all service information in another. Keeping your information together in one place is the preferred method, but whatever works for you as a practising manager is never the wrong way to do it.

You will have sustainability objectives, so try to make sure you keep as much of your information as possible in digital form. You will quickly find that your files become full of information relating to each building, with each building ending up with multiple files. Compiling your information in a logical way will help you to filter out duplicate information.

Understanding your service and how it fits onto your site

Everyone likes a sponge cake: two layers of sponge bonded together with a glorious cream. Facilities Management is a bit like a sponge cake. Your bottom layer of sponge is your building stock and the top layer is the services you provide. The glorious cream is your staff, who make the whole cake come together as one wonderful tasty treat. Perhaps a strange analogy, but if you look at the services you provide on the site as an overlay of the site itself, you will start to think holistically about how things need to be put together. One area that cannot be seen in isolation as it is part of the overall "jam sponge" is the FM Department.

Understanding the services you provide is a key and integral part of FM, ensuring that nothing is missed and that you stay compliant. List the services you provide and try and categorise them into two distinct areas: hard services and soft services.

Your hard services will be:

- fire and intruder alarm service and maintenance
- fire extinguisher service
- heating and ventilation maintenance
- fixed wiring testing
- PAT testing
- *Legionella* and asbestos control
- lift services

Your soft services will be:

- cleaning
- porterage
- minor maintenance
- security
- window cleaning
- grounds maintenance
- post

Note to self

Keep all the O & M information in one place; this makes it easy to access and update.

You may find that you outsource all your hard services and source all your soft services in-house. However, some soft services may not be practical to provide in-house, e.g. security. Listing your services will assist you in setting up your administration so that nothing is missed and everything is carried out in a logical way. Building up a picture of your entire site in this way makes sure you do not miss the essential cogs in the machinery.

Aide Memoire: The basics

- Get your administration structure right first.
- Make sure you have all the right equipment for the job: PC, phone, etc.
- Start to pull together a networking strategy; contact other similar institutions.
- Don't be afraid to steal other people's ideas. It can save you a lot of time in the long run.
- Stamp your authority by clearing out the old regime.
- Organise your information and administrative functions.
- Start to pull together an understanding of your site and property portfolio and how your service fits over this.

Chapter 3
Managing people: staff and customers

Managing your staff

Your staff are one of the deciding factors in whether your FM department is a success or failure. Success is fickle and can be destroyed by complacency. Your staff can make or break you, and finding the management styles that best suit the type of staff you have can make all the difference to your department. You will need to adopt more than one management style to deal with your staff, as no two people are the same and everyone needs to be treated in a slightly different way. In practical terms, adopting a rigid style to deal with all staff is doomed to failure. People are different, and what motivates one person in terms of incentives could be ineffective with another member of staff.

However, you must treat all your staff equally. Having favourites is a difficult issue. Sometimes you know who can and who cannot perform, and instinct and pressure of work dictate that you need to get the job done as quickly as possible. Inevitably, you may tend to favour one person, especially if you know they will come up with the goods. This in itself is not bad; you do, however, need all staff to be firing on all cylinders all of the time and you need to be careful. If you resort to favourites within your department, it will eventually come back to haunt you.

Staff need to fulfil a purpose. They need to drive your aspirations and ambitions and they need to fulfil the strategic objectives of your organisation. "Good staff are not born, they are made." To some degree, I agree with this statement. Your staff do not arrive on your doorstep fully trained and ready to be unleashed on your site. They are rarely completely up to speed with your philosophy on how your department should be run and neither are they always fully trained in terms of understanding the various theories of FM.

It is unusual for them to be self-driven, fully motivated individuals who will fulfil their role with the exuberance that you would like to see and with the efficiency your internal pressures demand. No member of staff is ever all of these things, although sometimes you may be lucky, and they are rarely so complete that you do not need to expend any energy in supervising them. Typically, you will need to fully train your staff to your way of thinking as early in the induction process as possible.

You need a five-point plan to ensure the success of your staff: choose your staff well; communicate with your staff; find out what makes them tick; "dangle the carrot"; and review.

Choose your staff well: recruitment

This seems logical; however, choosing them should be the biggest single investment in time that you give to your staff. These people ultimately ensure your success or failure. If you do not get the right person, you limit your ability to fulfil your overall desire for a perfect service. Ensure that you have the key ingredients before you begin the recruitment process.

Job description and person specification

Do you know exactly what you want the job to be? It is no good advertising for a plumber and then realising that you really need a heating engineer. Map out what you require the post-holder to do and be absolutely clear on the objectives of this post, e.g. how it will interface with other posts and what you want this post to be in one or five years' time. Cover all the existing tasks and responsibilities and clearly define what you expect of the role.

One of the keys to the success or failure of any department is ownership of the work involved. It may sound simplistic to say, but unless people can define their job and look to and for guidance in their job description, then you will always have issues with job content and ownership of the workload. Your job descriptions need to detail clearly and as unambiguously as possible the area of work to be fulfilled. Your job descriptions need to be looked at as a whole rather than position by position. You need to create overlaps to reduce your overall risk of items being missed. However, one word of caution: if you make your job descriptions too descriptive, you may have people picking holes in them with comments like "it's not in my job description." Your job descriptions should be worded in such a way as to give a simplistic overview and must contain the immortal words "and any other duties as reasonably requested by your manager."

All your job descriptions need to be consistent and uniform in their construction. You need to be sure that people on the same salary bands have commensurate job descriptions that are equal in content and cover the same level of work. This is really the subject for a full book on human resources. Making sure they are all equal across salary bands avoids divisions within the department and prevents accusations relating to fair and equal pay. This is particularly important in a large department. You may be lucky and never have to deal with fair and equal pay issues, but it could end up being your worst nightmare, affecting one group after another.

You need to be sure that this person is right for the role that you are advertising. Be clear about what skills you expect and what level of qualifications you require. Don't fall into the trap of thinking everyone should have a degree. If you set the bar too high, you may miss the perfect candidate. Also consider the rest of the team; you need a mix of talents and styles to make the team move forward in a positive way.

Interview techniques

Don't intimidate your candidates; after all, you are trying to get the right person for the job, not impress or overawe them. Make sure your questions cover the skills you want in your team. Make sure you also have a practical test of the required skills; this will speak volumes for the skill levels of the candidates taking the test. Don't rely on gut instinct or sheer desperation to fill a post. You will only make a rod for your own back further down the line.

Probationary period

Do not be afraid to use this. You need the right person and if you can't make that judgement in the first six months, then you are not doing your job properly. Set your new appointee tasks and push them to perform in the first six months. You will then know if they are capable of fulfilling the task well and this will also give you an indication of whether they can work well under pressure. Don't be afraid to get rid of them if they fail to meet the grade or do not show the positive progress you expect. This may sound harsh, but if you are not decisive about your staff during their probationary period, you lose the ability to get rid of them easily and make more work for yourself in the future by needing to continually pull them up for poor performance. You are also not being fair to the person you appointed if they are never going to make the grade. Remember, you cannot devote time to lost causes. You must also be aware that you are not treating your other staff fairly if you allow someone to “get away with it” and you are not being fair to the candidate who genuinely may not be up to it. Look after, nurture and encourage your staff as they are the key to success, not you.

Communication with your staff

Staff need to be part of every process that occurs in the successful FM department and communication with staff is by far the most important area you need to deal with. Most FM and estates departments will be responsible for a number of teams, such as Maintenance, Helpdesk, Site Supervision, Security and Cleaning. These teams need to operate independently, but are also an intrinsic part of the overall FM philosophy and must come together as one. Each of your teams needs to have a reporting structure. In general, this can simply be a meeting schedule.

What is a meeting?

> **Note to self**
>
> Set up a meeting schedule for all your team meetings, from Maintenance to Security; know exactly when meetings will take place and publicise the meeting schedule.

Meetings should be a general dialogue on the progress of the department. Sometimes meetings can be simply for the sake of having a meeting. This must never be so. A formal set of parameters should be drawn up, although these may change depending on the individual discipline. Your meeting structure should be relatively similar through each discipline. This is for a number of reasons.

- You are working with a team of individuals or team leaders within each department. Although each of these individuals will look to their own individual discipline, you must also encourage crossover between teams. This will leave you less vulnerable in the long run. If you have a uniform meeting structure, then the leaders in each team can easily pick up the thread of how each team is run.
- You yourself want to be able to pick out salient points within each set of minutes and a uniform structure enables you to focus in quickly on these. Remember what I said earlier: you will be swimming in a sea of information and anything that helps you get through that information quickly should be embraced.

- You must use your team meeting minutes in a practical way. They are a reporting and audit tool and you need to bear this in mind with every set of minutes you take. Information contained within these documents must be retained and held in a logical format as you may need them in the future.

Meetings are the best and the worst part of a busy FM department, but they are necessary and must be taken seriously, otherwise they become pointless.

A successful meeting should:

- *Be no more than 30 minutes long.* This means you have to get to the point and not waffle. Focus in on the essentials.
- *Be inclusive.* Everyone within the meeting must contribute. A "round-the-table" format works well and makes sure everyone has the opportunity to comment.
- *Be two-way.* It is not about people talking at each other. There needs to be a two-way flow of information.
- *Be informative.* What is the point of a meeting if there is nothing substantive to discuss?
- *Record actions.* This is your audit tool to gauge if your systems and strategies are working. You should be aiming to report your team's actions in a clear and unambiguous fashion. Your minutes should be a summary, but succinct and in a uniform format.
- *Develop strategy for the future.* Your meeting notes should not stop with you. They are a clear methodology for senior management to see if policies and procedures are working and if targets are being met. If you don't report this, how will they know if you are hitting your targets?

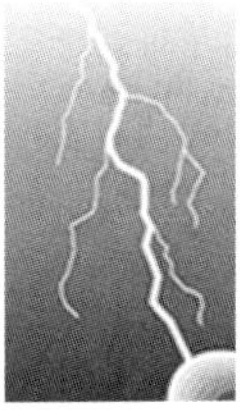

★★★★★★★★★★

Your meetings are the best way to get your message across. If the structure of the meeting is fixed and the minutes recorded, you have permanent record of your discussions and the timing of your requests is noted.

★★★★★★★★★★

Meetings can be a useful tool for diagnosing future risks to the organisation. If you leave your reporting to, say, a report once in every college term, you could be missing a valuable advance warning of risk to your senior managers and they will not appreciate that.

Meetings should be good vessels for two-way communication, but they don't always achieve this sufficiently for you to gain a full appreciation of what is going on in your department. Sometimes you get members of your team who are less comfortable with voicing concerns within a meeting format. You can't ignore these comments; you could miss essential elements of the FM process. You need to provide a forum for these comments to come through. This could be achieved by informal meetings; however, there is really no such thing as an informal meeting. If, during an informal meeting, something crops up that you need to action, inevitably you will need to make these actions formal. Your job as a manager is to manage. You cannot shirk your responsibility by ignoring comments. All this will do is alienate staff and create friction within your department. Ultimately, your staff will lose faith in your ability to deal with issues.

Therefore, even during informal meetings, make it clear that you will note items and that you may be forced to take action. There are ways and means of saying this in a diplomatic and supportive way so as not to put pressure on staff. There is a balance to be achieved. You cannot be completely black and white with no shades of grey. Your staff need to know you will be sympathetic and supportive. When you are rigid in your approach, you will promote the same response back from your staff.

The chat

Remember you are never off-duty and sometimes it pays dividends to just have chance chats with your staff. This will break down formal relationships and engenders trust from your staff (even if this is unfounded). You must be seen to be human. Try not to get over-friendly with individuals as this can be misconstrued as favouritism by other members of staff; you can't please everyone!

Suggestion boxes

> **Note to self**
>
> If you say you are going to do something, do it. If you don't, you will be seen as weak or ineffectual.

Suggestion boxes are a very useful method for capturing the things that you would sometimes miss, such as staff unrest or internal conflicts. Anonymous comments sometimes cut to the quick and can provide information that will help you to manage your staff and, if nothing else, you will get the occasional laugh from notes put in them.

Important: do what you say on the tin!

There is no point in having meetings, chats and suggestion boxes if you never action any of the items that are raised in them. Communication is not communication unless you do something with it. Always be true to your word and, if you promise things, keep your promises and, if you can't, explain fully why you can't. You need to build up a relationship with your staff and communication is the only way you can convey a message and then make sure that the message has been understood.

Theory and information slot

Maslow's Hierarchy of Needs specifies the following levels:

- *Physiological needs.* These are the basic requirements for human physical survival. These needs must be met first and include such basic essentials as food, water, shelter, oxygen and sleep.
- *Safety needs.* Once a person's basic physical needs are met, his or her needs for safety emerge.
- *Love and belonging needs.* Once safety needs are satisfied, the need for belonging and love emerges, including the need to belong to a group and to feel accepted.
- *Esteem needs.* These include both self-esteem and the esteem of others. Self-esteem includes the feeling that one is worthwhile. The focus is not on just surviving, but on doing well and improving oneself.
- *Self-actualization needs.* These are the needs associated with realising one's full potential; the person focuses on developing his or her talents and abilities to their fullest extent.

Adapted from Buchanan and Huczynski (2010), pp. 268–270.

Find out what makes them tick

I mentioned previously that each member of staff is different. In any team you will have a mixture of characters whose traits will all be different. What appeals to one person will not appeal to another. A job that one does too quickly, another one will struggle with. Take time to get to know what makes your staff tick.

Theory and information slot

Adapted from Buchanan, D.A. and Huczynski, A. (2010)

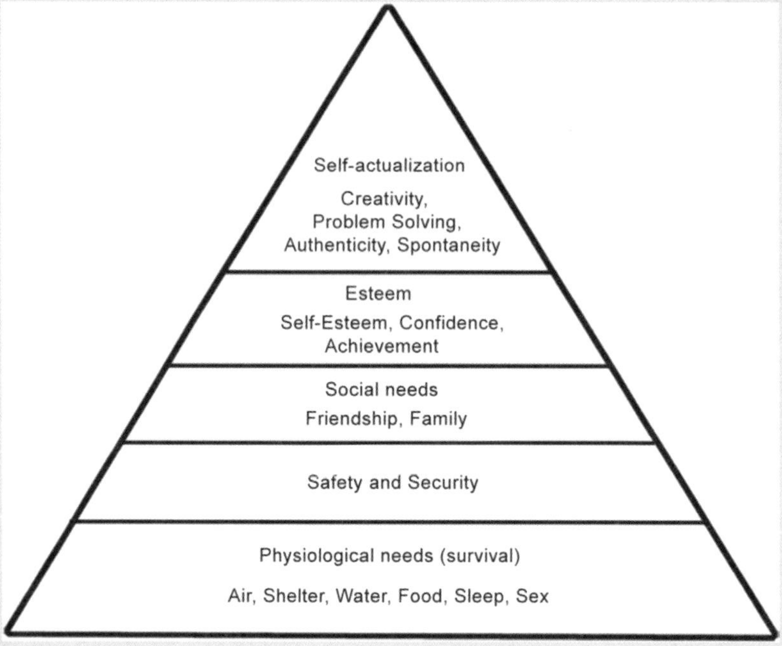

Don't be afraid to take notes (as long as you keep them securely); you can easily develop an overview of each member of staff, their likes and dislikes, strong points and weaknesses. These will help you to plan your training requirements. Once you understand your staff, you will get to know what motivates them and what their priorities in life are (these might not always be monetary). Getting the most out of your staff is about finding the right balance. It is about finding out what motivates them and, ultimately, if you can motivate your staff, you are halfway to winning the battle.

"Dangle the carrot"

The "carrot" can be anything you choose it to be. Very rarely are you, as a manager, able to throw money at issues and, likewise, you should never be throwing money at people to get them to meet your requirements. If it is contained within their job description, they should be doing it anyway. You will never influence some members of staff with anything but money; some will respond more favourably to things that create stability in their life, while other crave responsibility or simply want to do well and develop themselves. Your job is to find out what motivates them. Generally speaking, most of your staff will respond favourably to things like praise, competition, responsibility and CPD.

Praise

Praise is probably the easiest thing for you to provide to your staff; equally, it is easy to make it seem crass and meaningless. If you overdo praise, you will soon be seen as someone who is easily pleased or completely disingenuous. Praise does not always have to take the form of a pat on the back. Some people respond just as well to just simply being mentioned in a report or in the minutes of a meeting. When you are under pressure, this is probably the one area of staff management that you forget about. Give yourself a reality check and get back to being that supportive and praise-giving boss.

Competition

Staff cannot help themselves if they sometimes feel they are in competition with each other, particularly if this is seen as legitimate and fair. This is explained further in the theory and information slot on Hertzberg. Creating competition that is holistic and fair encourages staff. Again, caution must be your watchword to ensure competition does not create a culture of favourites or seem to favour one group over another, e.g. if you create competition in one department, you also need to do the same in the other departments.

Responsibility

People sometimes crave responsibility and this is an area where you can make your most dramatic advances with staff. Responsibility does not always have to go hand-in-hand with monetary compensation (see Hertzberg's "motivation factors").

It is practical to encourage people to take on additional responsibility to allow them to build up their skill levels within a department. You can create opportunity, even if there isn't any, by allowing people the chance to improve their skills while dangling the carrot of career development. You must, however, make it clear that although you are not able to provide a monetary incentive, you are able to provide support such as training and practical experience that could boost their confidence and their CV. You need to continually reinforce the point that you cannot provide monetary support, as what is seen as a perk in the first instance can be viewed by staff differently over time. If you do not continually reference the fact that you are helping their future employability and career development rather than working towards a pay rise, you will lose their flexibility. Discussing this with them will keep them on-board for longer.

Remember that you are not superhuman! You cannot take on everything and excel at it without help from your staff. This is the perfect example of how you can delegate tasks to individuals to bring them on. These are win-win situations because, not only are you building the skill levels of your staff, you are achieving aims that you yourself would not have had the time to undertake. You are providing yourself with a long-term continuity of staff as you have ready-made managers who can step into roles efficiently and quickly. You must, however, devote time to managing this process. If you simply delegate tasks to people without getting their original buy-in, you are likely to be seen as simply putting more work on one person. Make it clear that this is all about career development and opportunity for the future. Nurture this in your staff and be attentive about their progress and you will genuinely hit the jackpot. People appreciate honesty and if you are up-front about your intentions for them in the future, you can generally ask more of your staff.

Theory and information slot

Mary Parker Follett (1868–1933) described management as "the art of getting things done through people".

Adapted from http://en.wikipedia.org/wiki/Mary_Parker_Follett

Continuing professional development

This is a given criteria for any professional body. If an individual wants to further their career, part of that process is getting them to become members of professional bodies such as the Chartered Institute of Builders (CIOB) or the British Institute of Facilities Managers (BIFM). Once your staff join these professional bodies it becomes incumbent upon them to undertake CPD annually to keep current skills up to date and to move their skill levels on. If you encourage staff to become members of these institutions, then they will naturally want to move on and will be more likely to take on additional tasks, particularly in areas they are not familiar with, to enrich their own CPD and global experience. Getting staff used to the principle of CPD irrespective of whether or not they are in a professional institution is an excellent way of boosting skill levels and makes it simpler and easier to undertake your appraisal process.

Review

You have now put in place all the systems to make sure your staff are motivated and up-skilled to the point where you are likely to get the best out of them. It is now essential that you review what you have achieved, check that the skills and traits of your staff are the right mix for the services you provide and continue to look for ways to improve how you provide your service by adding new skills. Your review process should sit snugly into your QMS and should have the ability to move your team on progressively.

Don't forget to make your staff feel special. This is all about the personal touch. It is about making them feel like they are in charge, even when they are not. By "in charge," I mean in charge of their own destiny. If staff are aware that their own actions or inactions dictate how well or how badly they are viewed in the department, it helps them to understand how they can influence their own futures. Taking time to talk to staff on a one-to-one basis and offering praise makes them feel special. Joining in with staff events, even if this is the Friday night pint at the local, can help to build that bond between you and your staff. Again, you have to be aware of not creating favourites, so make sure you apply this principal evenly across all your staff. Remember also equality and diversity. If you all agree to go out for a pint on a Friday night, you could be excluding some of your team through religious or social beliefs. However, if you overdo this, it can be seen as insincere. It is all about getting a balance that works for you and feels natural.

What about you?

Never forget you are also a member of staff and you also need to be able to function. You need to be able to provide the service to the same level of excellence as your staff and you need to be able to respond to issues quickly and decisively. Three main areas you need to look at are: are you up to the job; do you have the necessary support structure; and are you adequately resourced?

Are you up to the job? The S curve and learning cycles

Ask yourself, do you have the necessary skills to be able to deal with the many facets of your job? Never be afraid to ask for help with something that you don't understand or an area that is unfamiliar to you. There is always the temptation when you are starting off as a Facilities Manager to say "yes" to everything blindly, without thinking through any long-term resourcing issues. Sometimes the previously mentioned sea of information can swamp

Note to self

Remember you are in a developing industry. However, the sheer diversity of functions allows innovation to drive the FM process forward.

you to such an extent that you simply drown in the sheer volume of work needed.

The S curve has been used for many years to define learning cycles. The S curve defines three phases. The first is the learning phase when growth is slow. This is usually followed by the growth phase, where growth is rapid as knowledge increases. This is followed by the decline phase when the learning and growth phases have reached maturity and

Theory and information slot

According to Hertzberg's: Motivational Theory, for a worker to be happy, and therefore productive, the overall work environment needs to be comfortable. Many managers believe that motivating employees requires giving rewards. Herzberg, however, believed that workers are motivated through feeling responsible for and connected to their work and contributing; in other words, making the work itself more rewarding. Hertzberg broke these factors down into two distinct groups:

- Hygiene factors. Aspects of work which remove dissatisfaction, but do not contribute to motivation, such as working conditions, salary, status, security and interpersonal relations.
- Motivation factors. Aspects of work which lead to high levels of satisfaction, such as achievement, achievement recognition, responsibility, advancement and growth.

The combination of hygiene and motivation factors can result in four conditions:

1. High hygiene/high motivation. The ideal situation, where employees are highly motivated and have few problems.
2. High hygiene/low motivation. Employees have few problems, but are not highly motivated.
3. Low hygiene/high motivation. Employees are motivated, but have a lot of problems.
4. Low hygiene/low motivation. The worst situation of unmotivated employees with lots of problems.

Adapted from Buchanan and Huczynski (2010), pp. 279–280.

additional innovation is needed to start the curve again. Sometimes as a Facilities Manager we could have two or three of these S curves going on at the same time and the sheer volume of information can derail your progress. This can then show as a marked decline in your productivity. It is the old adage of "Jack of all trades, master of none." The more you deal with, the less effective you become, until it gets to a tipping point where you actually start to undertake things badly. You become careless and mistakes start to creep in. Then, all of a sudden, the person who was considered a rising star becomes a liability and a poor worker. The trick is trying to see this point coming and off-load your unnecessary tasks in favour of concentrating on those that will pay the best dividends.

Your training

Don't forget you will also require training. You don't know everything and the speed at which FM grows means that you need to be taking on new challenges all the time and upgrading your skills to cope with the demands of new technology. Set yourself targets each year and keep your CPD up to the required level.

Do you have the necessary support structure?

Your support structure not only exists below you within your team, but above you at senior management level. It is important that you get these links right at an early stage to allow you to function correctly. Ensure that you get the correct level of support from your managers. Sometimes there can be a tendency to want to go it alone and only allow a light touch involvement from your senior managers. This can often be a mistake, as they are probably the best way of taking the pressure off you if they understand the problem. You need to convey these pressure issues to them effectively without it seeming like you are unable to cope.

Delegation can be one of the greatest tools for taking the pressure off you. However, this must be done wholeheartedly, while still allowing you to keep a firm hand on the reins. You have to make sure that your delegation is delegation and not abdication. You cannot shirk responsibility by simply giving a task to someone with the objective of blaming them if they fail.

Are you adequately resourced?

We live in a world where we are expected to do more for less. It is important to always keep a handle on what level of resource is needed for a task. Simply soaking up additional tasks without considering resources is a recipe for disaster. If you are under-resourced, you need to make that clear to your senior managers. If this affects your health, ultimately it is your responsibility to advise your senior managers that the volume of work cannot be undertaken with the available resources. You cannot blame them for putting you under pressure if you haven't advised them of a lack of resources. You owe it to yourself to avoid getting to the point where you simply cannot cope with any more work.

To deal with under-resourcing effectively you need to undertake a risk assessment of your current functions, identifying which of these are critical and which less so, but still need to be managed. When you reach the tipping point where you cannot undertake all your functions, your risk assessment will identify which areas are less critical. You can then report this to your senior manager as a defined risk to the department. This won't get rid of the problem, but at least you will have highlighted it. Remember that whatever you do, think of the process as a team function, think holistically about the task and the team together and remember that the best results will come from working towards a common goal as a team.

Teamwork: what makes a good team?

To drive any process forward within your FM department, you will need to create "the team," because you cannot do all this on your own. You need a positive contribution from the rest of your staff. So what are the basics of successful teamwork? Teamwork was explored by George Elton Mayo in the Hawthorne Works Experiment and his results can be summarised as follows.

- Does the team have clear goals and do they understand those goals only when they understand the defined outcomes necessary to truly focus on the mission? (see The Golden Thread in Chapter 4).
- Has the team created an environment in which people are comfortable to put their point of view forward and where members of the team trust each other and understand their role in driving the process forward?
- Does the team feel comfortable in being able to express potential solutions to issues, does knowing the way the team is set up allow them the freedom to express themselves without fear of being alienated or ignored by other members of the team, and does the team promote equality and diversity?
- Does the team have an identity that clearly defines them as a group and do members of that team feel as though they are a coherent part of that group? Have they developed a common goal that they all buy-in to?

Theory and information slot

George Elton Mayo was in charge of certain experiments on human behaviour carried out at the Hawthorne Works of the General Electric Company in Chicago between 1924 and 1927. His findings have contributed to organization development in terms of human relations and motivation theory.

Mayo concluded that:

- Work is a group and not an individual activity.
- The social world of the adult is primarily patterned around work activity.
- The need for recognition, security and a sense of belonging is sometimes more important in determining workers' morale and productivity than the physical conditions or environment they work in.
- A complaint is not necessarily borne out of factual issues; it is commonly a symptom of an issue with an individual's more basic or psychological problem.
- A worker's attitudes and effectiveness are influenced by both work and home demands.
- Informal groups within the work environment exercise strong social controls over the work habits and attitudes of the individual.
- Managers need to exact influence over informal groups to ensure cohesion within the work environment for the sake of the company.

Adapted from Buchanan and Huczynski (2010), pp. 305–308.

- Does the team consist of members with different but complementary skills which overlap and complement the skills of the other team members? Does the team have a mix of characters and skills?
- Is the team a place where aspiration, creativity and innovation, and different viewpoints are expected and encouraged, and a supportive, non-confrontational stance is seen as the norm?
- Can the team identify issues that may be stifling process and are they prepared to openly discuss those issues to improve progress while ensuring the team is not compromised by those issues or hindered in any way?

- Does the team have a set of guidelines or terms of reference to analyse issues that occur which also allow them to resolve those issues in a supportive way? Does the team have a collective set of values?
- Does the team have a leader who controls the processes while at the same time allowing input from individuals to amend and improve processes for the betterment of the group? Is the leader able to provide and maintain a clear view of the requirements of the group and guide the group while maintaining control?
- Are final decisions made collectively as a group and does the group as a whole have complete buy-in for those decisions?

Staff training

Note to self

Put together a full list of available training and construct a training matrix for the whole FM department. Include all the training needed to undertake each function

One of the most important issues in relation to your service provision is to have fully trained staff who can perform their duties to the maximum competency. Training is the key to this and fundamentally addresses the following issues:

- the ability for staff to undertake their duties effectively
- the fundamental need for people to want to progress
- the creation of stability and succession
- the ability to meet legislative requirements
- the ability to overlap services

Within the FM department there are naturally a number of different disciplines that staff work in and each discipline needs a different skill set to fulfil. No one can expect staff to come into the job fully trained and ready to go. Staff need to be developed and you need to take advantage of their willingness to learn. However, you will always have negative staff who fundamentally believe that they cannot be taught anything and, no matter how much you encourage them, they will never be able to grasp that it is for their own good.

Ability of staff to undertake their duties effectively

You carefully recruit what appear to be the best people for the job, but you don't spend time training them. Are they able to undertake their duties effectively? The answer is probably no. Even though you may have the right people and you may have clear procedures and service levels that are expected, they may still fail because they are using their interpretation of your procedures and this can produce a wide difference in service level. What you are looking for is consistency and the only way to get that is to ensure everyone gets the same message, told in the same way. Training will enable them to undertake duties effectively, but just as importantly it allows you to make judgements on performance knowing that each member of staff has received the same instruction. Without this overarching guidance you cannot expect staff to understand their role thoroughly.

People must want to progress

If you have recruited correctly, then you will have staff who want to progress. If you are recruiting people who don't want to learn and are not interested, then you have recruited wrongly. Most people have an in-built desire to improve themselves and will thrive on being set challenges that push them and allow the development of a potential career strategy. These challenges for career progression may not be the same for older staff members, but most people will rise to the challenge of learning something new to freshen up a job which may have become stale.

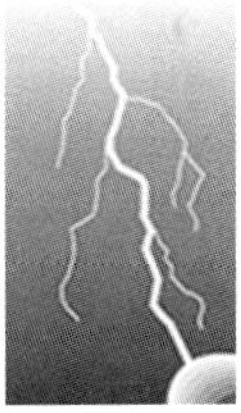

★★★★★★★★★★

Use your skilled staff to pass on some on-the-job training to other members of staff. Don't be afraid to tap into the skills of other people. You may be surprised at how well they rise to the challenge.

★★★★★★★★★★

When you have people who are not interested in training, it is far more difficult. You may take up a position where you have a considerable number of staff who have done the same job for a number of years and will no doubt say "you cannot teach me anything new" or "I'm too old to start learning new things." These are the people that you need to convince first that training, and retraining in some instances, will be good for them. Still not convinced? Well, neither am I, to be honest. Over my lifetime in FM I have met many people like this and I have always found it difficult, and sometimes impossible, to tackle people who are so set in their ways that they are effectively cruising towards retirement.

One way of tackling this issue is to appeal to their ego. Yes, these people may have an in-built distaste for learning something new, but if they have been in their job for a long time, they will have amassed a fantastic level of experience which you can tap into. Convince them that you need to pass on their skills to the younger workforce. Get them to construct a training schedule to pass on their experience. This could be as simple as, if they are a plumber, getting them to show all the staff where the cut-off valves are in each building and getting them to put that down on paper. Not only do you then get all your staff knowing how to shut down in an emergency, but you end up with a written record to update your O & M manuals.

You may be surprised to find that these staff then show a renewed vigour and may become more open to learning new things. Remember, everyone has a button you can push, although sometimes you may have to dig deep to find it.

Most staff will want to learn new things to provide them with opportunities. Your training must be controlled. Don't make everyone undertake the same training even if it is outside their job area. Set up a training matrix which shows you each person and all the available training, so you can start to plan and provide essential training in a controlled fashion.

Creation of stability and succession

Your training matrix can create a sense of stability in which all your staff feel confident that they have received the right training and they are confident they can perform their job to the right level and apply the right skills to the tasks involved. When people feel unsure about what it is they are supposed to be doing, it is likely to make them shy away from approaching things wholeheartedly and to not give their full potential as they may be afraid of failing. They will be more tentative in the way in which they approach the tasks as they may be unsure and looking for constant guidance as to whether they are doing things right. Being unsure in this way will make them less creative and unable to move outside the sphere of written instructions.

From a succession perspective, each time you recruit for a replacement staff member it is far easier to recruit from within. Ensuring that all your staff are trained in preparation for succession will engender a far greater willingness to take on extra duties as they know that in the long term they will be rewarded.

Succession and career progression are sometimes far more important tools than salary enhancements. If you show an interest in someone's future and help them in planning this, you immediately become more trusted in your decisions. From a perspective of change within your department, you are far more likely to be able to change processes and positions if all staff understand that they have a vested interest in that change and that you are training them to deal with it. Trust becomes embedded between you and your staff and they will, for the most part, understand that you do this in their best interests and in the best interests of the department.

Ability to meet legislative requirements

A lot of the work in a busy FM department revolves around compliance. To effectively deal with issues involving compliance, it is essential that staff are given full training on the implications of this and, in particular, what the effects on them and their employer are if compliance is not adhered to. Most of the work around compliance takes place in the background and is generally unseen. Inspections and testing, for example, are completely unseen until someone arrives to inspect your records. By that time, it is too late to worry. Training on legislative issues ensures that your staff are aware of their legal obligations and the consequences to both them and their employer in the event of inaction. Therefore training is emphatically the most important issue in relation to compliance.

Ability to overlap services

When you have specialisms within your department, you also become vulnerable from a number of different perspectives. Don't put all your eggs into one basket! Be aware of the following issues.

- If only one person knows how to undertake a particular task, how do you know if they are doing it right and at full efficiency?
- If someone leaves, it can leave you with a huge skills void which may be difficult to fill through standard recruitment.
- All your anecdotal knowledge around a subject could be tied up in one person, e.g. where keys are kept for machinery or stores.
- If someone goes on holiday or is sick, your service effectively stops.

It is important to protect your department by ensuring your structure allows the overlap of duties and rolls. This allows you to train other individuals to fulfil those tasks seamlessly in the event of sickness or holidays. An effective training matrix such as that given in Table 3.1 can identify those areas where you need to train others to do the same job.

How do you get staff to undertake training?

To some extent you will need to be dictatorial on how and when your training is to be undertaken. If you allow flexibility by allowing people to choose when they take training, this may mean that some of them miss out on training. Try to make training obligatory and set it up so that the whole department

Theory and information slot

Douglas McGregor was an American social psychologist who believed that there are basically two types of people in the work environment.

Theory X person

- An average theory X person dislikes work.
- They must be controlled and threatened to work effectively.
- They prefer to be directed and do not show initiative.
- Theory X managers do not give opportunities to staff as they wish to remain fully in control.

Theory Y person

- Physical and mental effort in a work environment come naturally to a theory Y person.
- They will direct themselves and are committed to the aims of the organisation.
- When the job is satisfying, they are fully committed to the organisation.
- They will seek responsibility.
- They have imagination, creativity and ingenuity.
- They are never fully utilised and are easily developed.

Adapted from http://en.wikipedia.org/wiki/Douglas_McGregor

can do it at the same time. When training involves legislative items it is clear that collective training can work effectively. Linking training to your appraisal system is also a good way of identifying your training requirements. It is also a good way to steer people in the direction that you want them to go by suggesting that they need to improve their individual skills via training.

Rewarding progress

In most cases you will not have the flexibility to give pay rises to reward progress. However, there are a number of ways in which you can entice people to undertake training other than threats. People generally respond well to the promise of long-term structured training, particularly if this can be done during their working day. Provision of vocational training to allow

Table 3.1 Example of a training matrix which can be adapted to suit your own organisation

Staff		Health and safety	Manual handling	Safe-guarding	Equality and diversity	Fire safety	Working at height	Lift release	PUWER	Asbestos awareness
Site supervisors	Employee A	M	M	M	M	M	R	M	R	R
	Employee B	M	M	Completed 12/10/12	M	M	R	M	R	R
	Employee C	M	M	Completed 12/10/12	M	M	R	M	R	R
	Employee D	M	M	M	M	M	R	M	R	R
Security	Employee E	M	M	M	M	M	O	M	O	O
	Employee F	M	M	M	M	M	O	M	O	O
Helpdesk staff	Employee G	M	M	M	M	M	O	O	O	O
Maintenance	Employee H	M	M	M	M	M	M	R	M	M
	Employee I	M	M	M	Completed 12/10/12	M	M	R	M	M
	Employee J	M	M	M	Completed 12/10/12	M	M	R	M	M
	Employee K	M	M	M	M	M	M	R	M	M
Administration staff	Employee L	M	M	M	M	M	O	O	O	O

(Continued)

Table 3.1 (*Continued*)

Staff		Health and safety	Manual handling	Safe-guarding	Equality and diversity	Fire safety	Working at height	Lift release	PUWER	Asbestos awareness
Cleaners	Employee M	M	M	M	M	M	R	O	R	O
	Employee N	M	Completed 12/10/12	M	M	M	R	O	R	O
	Employee O	M	M	M	M	M	R	O	R	O
	Employee P	M	M	M	M	M	R	O	R	O
	Employee Q	M	M	M	M	M	R	O	R	O
	Employee R	M	M	M	M	M	M	O	M	O
Environmental staff	Employee S	M	M	M	M	M	M	R	R	R
Grounds staff	Employee T	M	M	M	M	M	M	R	R	R
Supervisory	Employee U	M	M	M	M	M	M	R	R	R
Managers	Employee V	M	M	M	M	M	M	R	R	R
	Employee W	M	M	M	Completed 12/10/12	M	M	R	R	R

M = mandatory; R = recommended; and O = optional.

someone to progress in his or her career need not be expensive and, if you working in an education environment you can generally get your training at no cost. Encouraging people to grow, not only within your department but, more importantly, in the skills that will make them more employable elsewhere, will usually get a positive response. This is particularly true for staff who are young and at the beginning of their careers, or people who have become stale and want to learn new skills.

It may sound unwise to train people to give them the potential to move on, but investing in people shows them that you have faith in them. You take the chance that they may leave and take those skills with them, but sometimes you will find that being happy in the job is the prime objective of a lot of people. You spend a lot of time at work and, if you are not happy, you are constantly on the lookout for a means of escape from your current employment.

Sometimes people don't just crave additional skills to enable them to earn more money. Sometimes people crave power over and above money, and will take on additional training and responsibility for the opportunity to have some power. You must, however, be careful with these types of people as they can become more of a problem for you in the future if power goes to their heads. You can create monsters if you do not take the time to make them aware of the full picture and how they should behave with their staff.

Fair and equal pay and training

Warning
Ensure all your legislative requirements for the department are dealt with on your training matrix.

If you are offering training and responsibility in lieu of money, you must make clear the parameters of your offer and the reasons you are offering this opportunity. You must make it clear that this is not linked to any long-term strategy or potential pay increase. It is also a good idea to put these arrangements in writing to ensure that you are not challenged on fair and equal pay if someone is given responsibility that is comparable with a colleague who is earns more money. Care must be taken not to create ticking time bombs. Putting arrangements in writing removes any ambiguity and prevents a challenge in the future. Make it clear within any offer of training that this is not linked to pay and that it is there to allow the person more opportunity with succession or the external job market.

Customer service

Note to self

Make sure you budget for new uniforms throughout the FM department. Try to keep a common theme, but make sure that you do not exclude anyone in the department; go for practical designs.

Customer service is a key element in projecting a professional image while functioning effectively. What does customer service mean to an FM department? Usually, but not exclusively, the FM department is not seen as a front line service in the same way as front line reception staff. It is seen more as a cross-business service that is sometimes better hidden away. To be honest, who wants to see your painters, plumbers and joiners? In fact, you probably go to great lengths to make sure the service does its work without being seen. Sometimes this is for practical reasons, such as health and safety and to ensure access and egress. However, do not lose the opportunity to show off your service. You should be proud of the service you provide and should show how professionally it operates in its day-to-day activities.

Creating a good impression

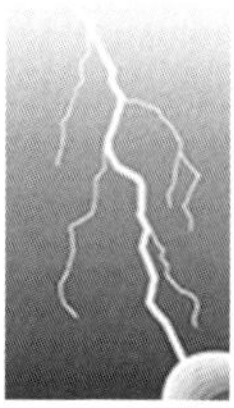

★★★★★★★★★★

When you choose your uniform, remember you need to put a logo on each piece of the workwear to avoid your staff having to pay for it through taxation as a perk.

★★★★★★★★★★

The first aspect of customer service success is to create a good impression. Let people know how good you are. Part of this process is the professional image of your department. Badge your services under a corporate umbrella. An FM uniform immediately gives the impression of a professional service. It also forms the other useful function of bringing your team together under one common purpose, providing your team with an identity that they can buy into.

Even if you have a corporate uniform for the FM department, changing it periodically helps to reinforce and freshen up that identity. For the new manager coming into a department, it will immediately stamp your ownership on the team and make them your team rather than your predecessor's team. It should never be an arbitrary decision to allow your team to choose the uniform. Plant the seed and let them come around to the decision you would have chosen. For practical reasons choose a serviceable colour which will last. White or bright colours will look good on day one, but will soon look shabby. Therefore take care to choose your colours and styles carefully. Do not buy cheap materials. They will soon look shabby and you will need to replace them early.

You will need to consider a corporate image. If the organisation has corporate colours or designs, it is appropriate to mirror these in the style and colour of the uniform you choose. You want your team to have an image which makes them stand out, but if that style grates or is completely different to the overall corporate image, staff will not feel part of the overall institution and this can be counterproductive. It may ultimately make the team look like a bought-in service and not an integral part of the business itself.

Providing good customer service

Good customer service has to be your ultimate goal. You can be as slick as you want in terms of provision of service, but if the customer does not like the service you provide, they can always exercise their right to look elsewhere. Without good customer service, your department could disappear or be outsourced altogether.

Good customer service is all about bringing customers back. It is about sending them away happy and providing you with positive feedback. Customers need to feel when they have received your service that they would be happy to use it again and become repeat customers. If you work to the premise that your customers won't have the opportunity to source your services elsewhere, you may never improve the service you provide as you do not have the incentive to improve. Always assume that the customer can buy your services in and then you will always be looking for a competitive edge to improve your customer services.

By definition, the FM department is continually shifting in terms of the services provided. New services are being added all the time. Providing a good customer service means your customers believe in you and that they trust you. It means that they look to you for solutions. It is all about building that relationship between you as service provider and the customer who uses your service. When your service naturally expands, your customer will have the confidence of knowing that you will deal with issues and deal with them in professional ways. The service you provide is ever-changing and developing. Your department needs to be able to adapt to suit this changing environment. If your systems are robust enough to monitor and manage the work, they will also provide you with the justification for new staff to take on new ventures.

Theory and information slot

We all strive to be truly great managers who earn the respect of many and are inspiring; in short, a pleasure to work for. But it is not easy and, as many of us know, great managers are few and far between.

So what are the qualities that make an effective manager?

1. *Integrity.* Integrity fosters trust, which in turn builds loyalty.
2. *Ability to empower staff.* A good manager empowers staff to perform at their best.
3. *Ability to motivate.* People perform best when they are happy and motivated.
4. *Ability to delegate.* A good manager understands the skills and abilities of staff and knows how to delegate.
5. *Adaptability.* Ability to adjust quickly and to understand the implications of changes, adjusting goals and strategies accordingly.
6. *Action.* To be prepared to take action and address problems as they arise.
7. *Role model.* Able to set an example to all staff of appropriate behaviour and performance standards.
8. *Value people.* Able to listen to people and be fair and understanding. However, this doesn't mean they can always keep everyone happy.

How do you develop, run and manage such a relationship with your client? You need to remember the one true secret of good customer service and acting accordingly: "You will be judged by what you do, not what you say." Good customer service will give you the edge you need and will develop a close bond between yourself, your department and your customers. After all, this is what we are fundamentally here to do: "facilitate." Providing good customer service is very simple. If you truly want to have good customer service, ensure that you carry out the following suggestions.

Answer the phones

Make sure that someone answers the phone when it rings. Your customer is waiting to give you a job; this is what you are here for and if you don't answer the phone, you have failed and fallen at the first hurdle. Make sure your Helpdesk phone is always answered and set up call groups where the phone will divert to if it is not answered within the first six rings. Make sure that your staff have all the tools and training they need in terms of telephone answering skills. Time taken working on these skills provides your department with a professional image and, if you're not endeavouring to be professional, why are you here?

Don't make promises you can't keep

You should not plan to keep your promises: you must keep them. That's what a promise is. Reliability is one of the keys to any good relationship and good customer service is no exception. If you say you will provide delivery today, then make sure it is there on time. If there is a repair to be done and you say it will be done tomorrow, do it today. Set yourself challenges to under-promise and over-perform. If you don't and the opposite is true, you will always be seen as a department which fails to come up to the mark. Think before you give any promise, because nothing annoys customers more than a broken promise.

Listen to your customers

Is there anything more annoying than getting a call back from a contractor or supplier saying: "I didn't quite catch what you said, what time do you want the service?" or "What was your name?" Listen to your customer. There is no magic formula, but you should encourage your staff to take notes during phone calls. Your Helpdesk is probably the hub of your department, so make sure these basic principles are engendered in this team first. Be ready with a pen as soon as the phone rings. One of the most satisfying things when dealing with someone over the phone is when, at the end of the call, they are able to remember your name, which was probably said only once at the beginning of the call. It makes you feel wanted and special, which is what good customer service is all about.

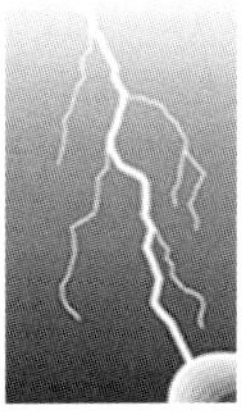

★★★★★★★★★★

Make sure you deal with complaints in a formal way. Never allow them to be forgotten or swept under the carpet as they will come back to haunt you at a later date.

★★★★★★★★★★

Deal with complaints

No one likes hearing complaints and many of us have developed a reflex shrug saying, "You can't please all the people all the time." Maybe not, but if you don't try and don't have a philosophy of the customer always being right, you will never get it right. Give complaints your full attention and set yourself deadlines for responding to them. Include dealing with complaints as an internal SLA so that you can monitor your completion rate for dealing with them. If you give the complaint your full attention, you may be able to please this one person, this one time. You will then inspire the confidence for this person to trust you in the future.

Be helpful

> **Note to self**
>
> Pull together a training programme for all staff. Apply this consistently across all departments so everyone is singing from the same hymn sheet.

Sometimes people will ring the FM department knowing full well that you will not be able to help them with a direct service. Engender your staff to try and help, to try and point the client in the right direction. Develop a name as the department which can and will go the extra mile to please the customer. Never moan or grumble about customers wasting your time. Never forget that good customer service is what your department is about.

Train your staff to be courteous and knowledgeable

Do the training yourself or hire someone in to train them. Most importantly, give every member of your staff enough information and power to make those small customer-pleasing decisions, so he or she never has to say: "I don't know, but so-and-so will be back at...". Again, this is ultimately down to you. If you don't engender this type of behaviour in your staff and department, and expect it from them on every occasion, then you are in the wrong job.

Go that extra mile

> **Note to self**
>
> Make sure you yourself are up to speed on customer service before you roll out any training programme across your department.

Go that extra mile. If someone asks the direction to somewhere on the site, show them on a plan of the site or even get someone to take them to the place. When someone shows up on site and asks where they can park, don't just say the car parks are full. Have some answers ready about local parking and fees. People don't like to feel abandoned. They may not say so to you, but people notice when you make an extra effort and will tell other people.

Throw in something extra

Customers appreciate getting something for nothing, especially when they don't expect it. If some flat-pack furniture is delivered to the site and a particular department wants it delivered directly to them, offer to put it together for them as well. Add-ons or that little extra finesse to the service will get staff coming back to you for more. You cannot be expected to do everything, however. Don't get drawn into doing things that are patently outside your scope of delivery, unless there is some genuine reason or advantage in doing so.

Practising what you preach

Your department is the all-singing, all-dancing department which goes that extra mile for the customer. You are getting fantastic comments back and your customer service is deemed a success. Don't be the weak link in the chain by not practising what you preach. As the figurehead of your department, if you don't provide the same level of service as your staff, you may undermine the overall effect. If you have extolled these virtues to your staff, but don't come up to the same level yourself, how does this look to your customer and to your staff?

Make sure that you follow the same basic principles and engage with your customers. You have a vested interest in having close links with them. You will always get an honest answer from other department managers if they feel you are interested in what your department's performance is like. Arranging regular meetings with your departmental heads (each term in an education environment) will pay long-term dividends for you. You will pick up poor performance early, allowing quick and decisive reactions and further enhancing your service. You will also start to develop a bond with your customers. It is all about loyalty. If your customer feels you do everything possible for them, then they will always be your best ally when, inevitably, things go wrong. Never underestimate the potential power of these close relationships.

Development of customer services

Customer service is not something which is put into place and never looked at again. It is a process of continuous development in which service levels are moved onto the next level of expectation. Working closely with your customers allows you to gauge what their perceptions of your customer service are and whether you meet the expected level of service requirement. You can only do this through SLAs, which are dealt with in detail later. Your team should always be looking to improve your service. In this way you will always have the competitive edge if your service is market tested against external organisations. In a college environment this is dealt with by the SAR (Self-Assessment Reports) process.

What is a SAR?

In real terms, a SAR is an annual report detailing what your service intentions are for the year, including the allocation of potential resources, the setting of time-frames and naming responsible persons. This SAR is then monitored as the year progresses and is usually "ragged" (red, not complete; amber, partially completed; and green, completed) in terms of completion status. Arguably, your SAR should contain all your SLAs, but in real terms this would make them too unwieldy to monitor. SARs generally deal with significant functions, whereas SLAs may deal with the minutia of the service provision. In either case, your SARs and SLAs must be your starting point. The principal of a SAR can be used in any FM department and is not exclusive to an educational environment. An example of a SAR is given in the Appendices.

Warning
Do not push service levels up without considering the implications for your budget (don't promise things you can't resource).

You should review your service standards in discussions with your customers and suggest ways in which they can be improved and incorporated

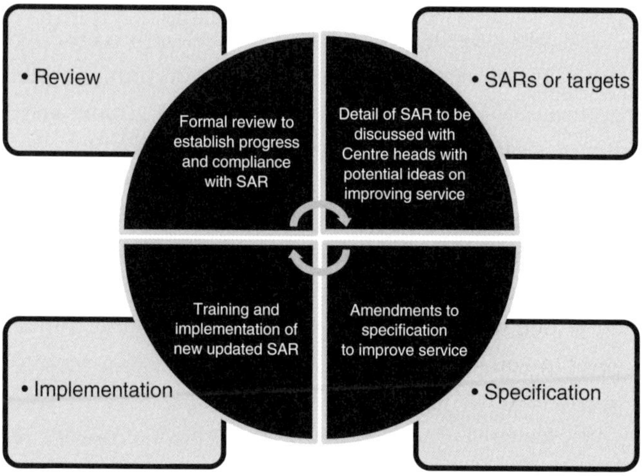

The process of a SAR in action and how it creates a virtuous circle

into new and updated SLAs. As ever, this must take into consideration budgetary requirements. Never improve service quality without taking into account any budgetary effects.

Reviews of this type follow a set pattern and can be mapped out fairly easily by: discussions with department heads; altering specifications to suit changes; implementation; and review of progress (see diagram). The whole process is cyclical and, in real terms, never ends as you strive for continual improvement of your service.

Your customers are not just your department heads, however, and to discount everyone else's views would be folly. Therefore you must engage with all your customers and treat every contact in the outside world as a potential customer.

Students

In an education environment, the students are the people that we are here to provide the service to. Without them, we wouldn't have a service to provide. Your students will not shy away from giving you their views and engaging with students is critical to your success.

There are a number of ways the FM team can obtain feedback from students. The first is by attending student meetings. Most curriculum centres will be happy to allow you to attend student meetings. This is your chance to come face-to-face with your customers. This can be quite daunting, as you can never be sure what is going to come up at the meeting and what you are going to have to answer questions about. However, if FM is flagged up as a regular slot on the agenda, then you may get some very useful feedback on what the students are looking for from your department. Do not, however, ignore comments from a student meeting. You must respond positively to requests and not brush them under the carpet, no matter how trivial they may seem.

The QDP is a national bench-mark of how students feel about certain aspects of college life. This is usually curriculum-orientated, but there are a few topic headings that are useful to the FM team, such as: does the college feel safe; are the toilets clean; and does the college feel warm and welcoming? All of these will help to indicate where you need to improve. Use a weighted scoring system which will help you see your performance against the previous year. This will help you to target where you need to change in terms of next years' service and also help you bench-mark against peer organisations.

Customer service cards are a good way of engaging in a non-contact way. You can strategically leave feedback forms for students to complete in, for example, catering areas. These could include questions such as: have you ever used the services of the FM department and, if so, was your experience good, average, poor; are the common rooms clean; or are the toilets clean? All of these will help you to gauge what it is that the students are looking for from the FM department when they come onto the site. Never assume that the student is just looking for qualifications. Increasingly, it is becoming all about the "learner journey," the experience that the student has when they come to college. From the moment they come through the door of your college to the time they leave, they are picking up experience of the real world and college is the transitional step between school and the work environment. The experience they have will encourage them to come back time and time again to improve their individual CPD.

Theory and information slot

One of those credited for the invention and development of the SWOT analysis is Albert S. Humphrey.

A SWOT analysis looks at:

- *Strengths*. What are we good at?
- *Weaknesses*. What are we bad at?
- *Opportunities*. What are the chances of improving performance?
- *Threats*. What could cause us problems?

Adapted from http://en.wikipedia.org/wiki/SWOT_analysis

Aide Memoire: Managing staff and customers

- Set up a communications strategy for your whole department specifying meeting schedules and strategies.
- Check available skills and evaluate the skill requirements of the department.
- Construct a training schedule for the whole department, making sure you consider the strategic objectives of the organisation.
- Develop customer service for each section, making sure your staff focus on customer service techniques.
- Listen to your customers.

Chapter 4
Policies, strategies, procedures, risks and liabilities

Your policies and procedures are the key to everything you do within an active Facilities Department. They define what you do, why you do it, and how you do it. Your policies and procedures will be dictated primarily by legislation and the strategic objectives of your organisation. In general, policies are generated by legislation and have legal implications. Similarly the risks and liabilities of your building stock and staff need to be interwoven and covered by your policies and procedures to ensure your compliance is complete and thorough.

Legislation can be simple and straightforward in that the requirements are laid out and it is usually fairly easy for you to determine the salient points. Every business will have in place a raft of policies to ensure compliance; however, compliance can change and you will generally find that some policies are put together in response to a particular piece of legislation. These are committed to paper, communicated to the staff who are trained in the appropriate area, then they generally sit on the intranet awaiting future use. Sometimes slight variations in legislation and updates can be missed and unless your Quality Assurance (QA) system picks these up (as it should), then policies can become out of date. Your first objective is to ensure that any policies are current in their application and meet the organisation's objectives as well as all the necessary legal requirements.

The Appendices to this book give a number of the policies you may need to comply with. You can add further policies to these examples to meet the aims of your own organisation.

Policies

Policies are not just the pieces of paper required to ensure compliance; they act as definitions so that you and your staff can improve your department's performance. You must ensure that all of your policies have the following characteristics.

- *Thorough.* Loose or open policies cause confusion. When something cannot be clearly defined, it cannot be followed clearly.
- *Incorporate objectives.* Your organisation must have strategic objectives for its long-term survival and these must be incorporated into your procedures. If they are in your procedures, they also need to be in your policies; if your policies do not mention strategic objectives, how will they be measurable or dealt with at a corporate level? Your strategic objectives must be incorporated into your policies otherwise they serve no useful purpose. Most policies deal with compliance in some way. For an organisation, compliance is legally binding. Your strategic objectives will include risk aversion, which includes legislative requirements.
- *Applied.* Policies can be rigid definitions. Your policies and procedures will form part of your training objectives and will drive measurable targets.
- *Trained.* Never just give policies to staff and expect them to simply run with them. You must be prepared to invest time in making sure that the fundamental messages within the policies actually get through and you cannot do this without using those policies as part of your staff training. This also helps to ensure that your policies are robust and uncompromising in terms of your strategic objectives. Training often identifies loopholes or missing information within your policies and is a way of reviewing whether they actually work.
- *Review.* Nothing stands still in FM. The goalposts are continually moving and your policies must also be capable of moving with your department's development. You must have in place a formal review of each procedure and policy to keep track of changes. This can be formally built into any QA procedure such as ISO 9001, which we will deal with in a later chapter.
- *Simple.* Policies in particular should be a framework for your procedures to be built upon. If they are long winded, you may never pick out the embedded message. Try to keep policies to one page if you can. If you get into this habit you will be forcing yourself to get the message over in the least number of words. Your procedures can add detail to the policy.

Strategies

> **Note to self**
>
> Your strategies will need love, care and attention. They need to be changed, updated, monitored and generally maintained to ensure they work for the department.

To move your department forward in a positive way, you need to define what you want to achieve. Creating your strategies is the first step in defining the service you are trying to apply. Clearly setting out your strategy helps you to focus on the service you are trying to provide and to define the expectations of the users of the service.

Strategies describe how you are going to turn the aims and aspirations of your policies into action. They define where you hope to be at a certain time and are your route map of how you are going to achieve the targets you have set yourself. Your strategies should have the following characteristics.

- *Time-bound.* They should define when you are going to achieve your aims.
- *Defined.* They should start to put some flesh on the bones of your policies in terms of specific areas that you are going to look at.
- *Realistic.* They should represent an achievable level of service, while pushing the boundaries of that service to ensure improvement.

What is a strategy?

A strategy in the simplest terms is applying the parameters of the service you will provide in a clear unequivocal format. It is your plan of action and details fully your intentions.

This is a practical guide, so I am not going to concern myself with the theoretical ways in which you can analyse your processes to develop your strategy. In the end you will be concerned with practical issues and the correct strategy is whatever works for you. There is no "one size fits all" solution.

Never fall into the trap of creating strategies and then sticking too rigidly to them. Your strategies are a working open document which should be able to change when required. You need a framework and it has to be clear and concise, but the work of a Facilities Manager is always developing and you need to be flexible. Changes in our business may be forced by legislative, operational, or technical requirements and flexibility is essential.

Strategy is about setting a firm set of parameters, but allowing fluid movement of those parameters to meet the changing requirements of the site. An example is heating strategy.

- *Who?* Whose responsibility is it within your team to monitor and manage this strategy and what are their individual responsibilities on a day-to-day basis? Clearly defined lines of responsibility will not only help to clarify who should be dealing with this item on a day-to-day basis, but will also allow you to interface this strategy with key objectives and link into the appraisal process to make appraisal targets easy to construct, manage and monitor. Your strategy should contain a flow chart indicating the lines of interpretation of the strategy to help provide a clear and unambiguous document.
- *What?* For example, define what heating is. Is this permanent or temporary heating? Is it linked to any other systems such as variable refrigerant flow (VRF) or ventilation strategies? Is it controlled by a BMS and, if so, what are the parameters by which the BMS will control it, e.g. set points, eco-settings, optimisation and compensation? What are your targets? Do you aim to keep the heating between 18 and 21°C? Will you aim to hit a constant 18°C 30 minutes before the first use of the day?
- *Where?* Where are you going to be heating and how is that influenced by the occupant's use of the space, e.g. bakeries, workshops or art studios.
- *When?* Define when you are going to heat the building, e.g. will there be a defined heating season? Engage with users to ensure the provision of heating matches the work provision. There is no point heating a space when there is no one in the building and, conversely, you need the heating at an appropriate setting when the building or space is occupied.

Note to self

Make a list of all the areas within your department where you will need a strategy. You don't need to draft these immediately. Take time to get the job right.

Your list of strategies for your busy FM department may include the following: heating and ventilation; opening and closing; PPM; responsive maintenance; and Helpdesk.

What are the key points of a strategy?

Your strategy is your guidance document for whatever subject you are trying to define. To be effective, it has to be the following.

- *Workable.* There is no point in creating a strategy that is too difficult to manage or work. Do not create work for yourself by having content that is unrealistic or unworkable. Your strategy needs to be a practical working document and brevity should be the order of the day.

- *Achievable.* Your targets need to be realistic and achievable; there is nothing more demoralising to staff than to be given targets that they cannot achieve. Set your targets modestly in the first instance and allow for over-performance. This will spur on the staff who have to work with this document and will give results when applied to your appraisal targets. Your document should be a live, working, dynamic document capable of changing with the changing requirements of your organisation. If this is the case, then you can slowly increase your targets each year, allowing yourself time to build up experience and technical knowledge while, at the same time, allowing your department to perform and show continual improvement.
- *Inclusive.* Your targets must encompass the wider community. Sometimes your ideas may not take account of the technical knowledge of your staff. Allowing them to contribute will let you tap into their skills and expertise and also give them ownership of the strategy. Do not forget the expectations of the end-user. Allow them to set some of the parameters for you, even if these are unrealistic. Sometimes they may ask for things that are in conflict with other priorities, e.g. a user may ask for internal temperatures to be at a constant 23°C. You then have the opportunity to discuss your sustainability agenda with the user. Without explaining decisions, your objectives may not be fully understood and, if the users are not fully on board, you risk constant complaints such as "My workplace is too cold."

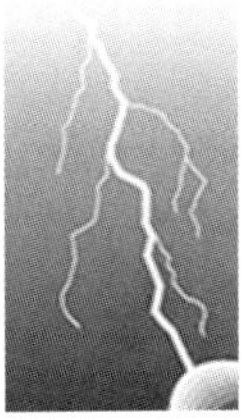

★★★★★★★★★★

Make life easy for yourself and use the templates in the Appendices to construct your policies.

★★★★★★★★★★

- *Comprehensive.* You will never tick all the boxes because the goalposts are continually moving. If your strategy is fluid enough to absorb these changes and is maintained as a live document, then it will be

Note to self

Start to pull together a full list of all your FM policies, strategies, procedures and forms.

as near to comprehensive as possible. As with any live document, you need to maintain and regularly update it.

- *Fluid.* All your documents need to be able to quickly adapt to changes without compromising their overall effectiveness.

Strategies are part of your tool kit for explaining the service you are providing and, like any tool kit need to be maintained. It is useful to link all of your policies, procedures and strategies to your QMS. Sample policies, procedures and strategies are given in the Appendices.

Procedures

Your procedures are the nuts and bolts of how you will carry out your policies. They define how you will achieve the goals and aspirations set out in your policies.

Procedures should have the following characteristics.

- *Purposeful.* There is no point in writing a procedure that does not have a purpose. Everything you do within your busy department must have purpose and a reason for doing it. Some things can be difficult to quantify, but they must all have a purpose.
- *Linked.* Procedures must be linked to specific policy documents. Your policies should drive your strategic objectives and, if your procedures are not linked or annexed to a policy document, then they will not be fulfilling any of your strategic objectives.
- *Specific.* Procedures should specify the level of service you are trying to achieve and how you are going to achieve it.
- *Owned.* Your procedures must be aimed at a particular individual or group of individuals whose duty it is to action them. There is no point in having a procedure that does not have a target owner to action it.

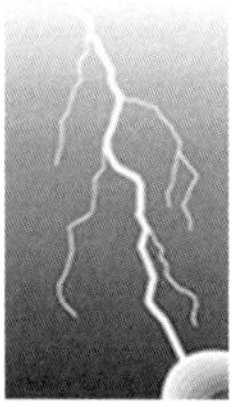

★★★★★★★★★★

Try to get a common feel to all your policies, strategies, procedures and forms. Connecting them in this way helps the flow of the documentation.

★★★★★★★★★★

- *Measurable.* If you have written policies and procedures and your staff have complied with them, but you are then unable to measure the effectiveness of the whole process, then there was no point in developing policies and procedures in the first place. If you do not build measurability into everything that you do, your service can never be measured or shown to have improved.
- *Lean.* Your procedures should be as lean as practical. Clarity of message comes from keeping things simple.

Guidance notes

From your raft of policies and procedures, you will have derived a number of tasks required to fulfil your policies and procedures. Both policies and procedure should be kept brief. However, brevity sometimes misses out detail. This is where your guidance notes come in. Guidance notes explain processes and practices in greater detail. They allow you to be more specific in your procedures. Guidance notes should have the following characteristics.

- *Specific.* Your guidance notes are your last chance to explain a process to your staff before they begin to action it. The level of detail will depend on the complexity of the process and what you are trying to achieve.
- *Bite-sized.* Policies and procedures provide an overview of what can sometimes be a complex subject, with many facets of the service provided included in each procedure. Your guidance notes should focus on each individual facet of the service and define what is required.
- *Able to be monitored.* Your supervisory staff should be able to use these guidance notes as a monitoring document to ensure that staff are complying with your requirements.

Theory and information slot

Strategic management is developing plans, policies and procedures that drive the organisation's strategic objectives and mission statement forward. The process can be measured in many ways. Modern establishments sometimes use a "dashboard." This was developed in the 1960s by French process engineers who created "tableau de bord" literally "dashboard."

- *Readable.* Avoid the use of esoteric language and keep the text as simple as possible.
- *Useable.* Guidance notes need to be practical documents that are readily understood and capable of being used on a day-to-day basis and as a training aid.

Forms

Forms have been much maligned over the years as a bureaucratic embellishment for the paper-pusher. However, a form is your auditable compliance document. It will show that you have complied with a particular task while providing your audit team with written proof of compliance.

Forms should have the following characteristics.

- *Simple.* Keep your forms as simple as possible. Simple tick-sheets are far easier to fill in than long-winded handwritten descriptors of finished jobs. Don't forget that you have to audit these sheets; simple ticks are easier to check than reams of handwritten notes.
- *Concise.* Keep the information on each form to a minimum. This will make it easier for you to audit, easier for your staff to use and easier for them to be trained to use it in the first place.
- *Time-bound.* Don't forget to put the date forms. Simple as it may seem, if a form is not time-bound or time-stamped, then it really does not serve much purpose as an audit document.

Administrative flow

Try to create a natural flow of documentation that will assist, not hinder, you in meeting compliance obligations. An example of this is given in Appendix 1 and in the diagram below.

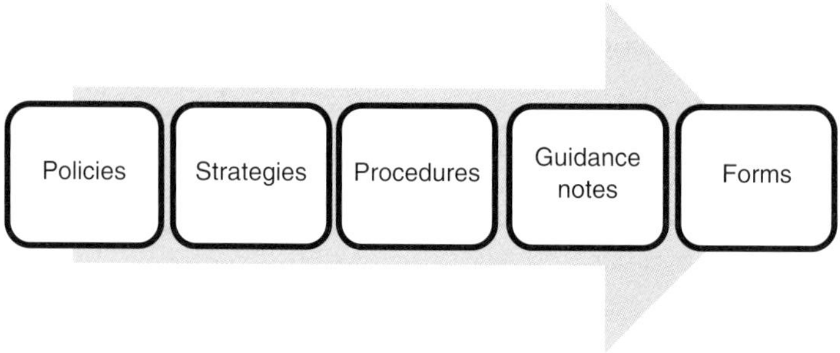

Theory and information slot

Policies, rules and procedures are fundamental to an organisation. Rules guide employees' actions, enabling consistent and acceptable behaviour and establishing prescribed levels of performance.

Adapted from Hellriegel, Don and Slocum, John W, Addison-Wesley Pub. Co. (Reading, Mass.): 1978, 2nd edition.

Ten reasons to use policies and procedures:

1. Allows a framework for your QMS.
2. Clearly defines and interprets strategic objectives of the organisation.
3. Clearly defines who is responsible for what.
4. Helps to link together various disciplines and subject areas.
5. Allows you to track work when linked to strategic aims and objectives.
6. Can provide a logical flow.
7. Reduces confusion amongst the workforce.
8. Ensures proper record keeping.
9. Ensures specific legislative items are contained within all spheres of work such as health and safety corporate social responsibility, equality and diversity.
10. Defines clear management and control.

Monitoring and managing

Let's start with the principles of what is monitoring and what is managing. Monitoring is the process by which you control an activity by proactively following its progress using key intervals to bench-mark the success or failure against a set of criteria or performance indicators. Managing in the context of this chapter means maximising the available resources to enable effective monitoring to effectively control the process in question. As this is meant to be a practical book on how to run an FM section from scratch, I hope by the end of this chapter to have shown you the value of analysing what you are trying to manage to maximise your time as a Facilities Manager.

Monitoring and managing can apply to monitoring and managing the following: staff, contracts, resources and environments. Although the principles for each of these areas will be similar, they all vary in subtle ways and will be dealt with separately. The areas you need to focus on are:

- What are you managing and monitoring?
- Where are you are managing and monitoring?
- When will you manage and monitor?
- How will you manage and monitor?
- Why are you managing and monitoring?

Each of these areas needs to be defined to help you focus on the process so that you do not miss anything fundamental.

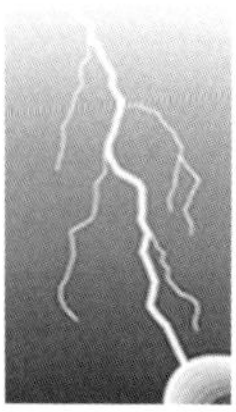

★★★★★★★★★★

Pull together a matrix for all your contracts, pinpointing where and when each service is provided. This will give you an overview and allow you to see where your resources need to be focused.

★★★★★★★★★★

What are you managing and monitoring?

Unless you know exactly what it is you are dealing with, how you can be 100% sure you are effective? You need to define the process or activity you are managing and a set of parameters for the service or service levels. For example, if you are managing a cleaning contract, do you have a specification that defines what is to be cleaned and to what level? If you don't have this simple information, how can you control the contract? If your specification is prescriptive in its wording, it is easy to check if it is not being carried out.

Where are you are managing and monitoring?

Note to self

A bit of careful analysis will help you to understand what you are monitoring and managing and why you are monitoring and managing it!

Do you know where these services are being provided? Is it clear from the descriptions in the contractual documents where the services are provided? If not, then this can lead to interpretation and, wherever there is interpretation there can be subjectivity. This can lead to unclear instructions, which can lead to increased costs. It is equally important to make these distinctions with in-house contracts because this can also lead to subjectivity. When you are unable to define poor performance, you are equally unable to correct poor performance, and without that ability you are unable to control.

When will you manage and monitor?

There is no point in having a watertight contract, specification or service level agreement if you do not define a frequency at which you need to monitor and manage. It is far easier to manage something if you set aside time to do so. Make sure you book in, and adhere to, periodic meetings on progress. This shows the contractor two things: that you are methodical and that you are checking the contract. Without this, your contractor will take advantage of you. They will assume you are not bothered about quality and that you are easy to manipulate.

The advantage of a formal meeting is that you have an agenda and a methodology for tracking changes to the contract and a diary of discussions, which helps you to control costs and variations. Put all your checks on quality in a diary so that you don't forget to do them, including all your key dates such as monthly valuations and contract start and end dates. This is a simple task in Outlook and ensures you are ready for re-tender in good time with the option to formally review how well the previous contract has gone. Also put formal meetings with in-house staff in the diary to look at progress and to check you are still following service level agreements and that the service is not suffering from drift. Any formal checks or audits that you are going to do should also be in your diary. This ensures that you make time and space for these meetings. If you don't, something else will come in to fill the void. We are sometimes our own worst enemy in terms of filling our diaries to make sure that we look busy to the outside world. We then somehow convince ourselves that these tasks have enough meaning and substance to be left in our diaries.

> **Note to self**
>
> Use your Outlook calendar to put in all of your management activities. This ensures you find time to monitor issues prior to a significant event such as a valuation or on site meeting.

Your job is to manage contracts and you never stop managing, from the moment you walk through the door in the morning to when you walk off-site; sometimes it does not even stop there. If you want to be successful in FM, you must never ease up on management. Never assume something is running smoothly and does not need to be managed. You will never be in this position, because the moment you relax your hold, your hard work will unravel.

How will you manage and monitor?

There are basically three ways in which you can manage and monitor a contract.

- Get out there and look at the service at ground level and check that it is being done to specification and that it is reaching the desired service level. Go out and muck in with the workers to test out assumptions on service levels at grass root level. This, for me, is the best and most honest and up front way of managing as it shows the contractor or your staff that you are not afraid of getting your hands dirty. You will gain so much respect from operating in this way that it is well worth the effort.
- Get the contractor to justify the contract success by providing you with raw data such as signing-in and signing-out sheets, worksheets and customer care questionnaires. This is particularly good where you are under-resourced yourself. It can, however, allow the contractor to manipulate data to give a better outcome. Don't forget the old adage that there are lies, damn lies and statistics!
- The third method is to employ someone else to monitor the contract for you. This is particularly useful if you are under-resourced in terms of time and if there is a specific technical requirement to the contract which is outside of your skill set, such as lifts.

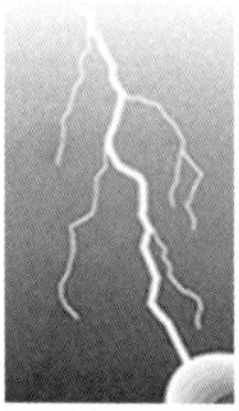

★★★★★★★★★★

Showing your staff that you are capable of doing the job by rolling up your sleeves and helping out will improve your credibility and help to create a bond between you and your staff.

★★★★★★★★★★

Whichever way you choose to manage and monitor, make sure you build in quality checks as you go. We have already discussed appraisal processes as a methodology of pushing forward quality by identifying poor practices and improving them. You must not forget to apply those principles to yourself. Not only is it likely that you will be appraised yourself and may wish to prove in statistically how well you are doing, but it should be standard practice for you to take a step back and apply a bit of self-analysis to your own performance. Building in quality systems to provide you with statistical data to allow you to analyse your performance can be an eye-opener. If you receive performance-related pay, this is essential.

Why are you managing and monitoring?

Do you have a clear mandate on what, when and why you are managing and monitoring something? Are you monitoring something that does not need to be monitored? As with everything in FM, we can sometimes do something just because we think it is the right thing to do, because it is there and we feel we need to control it. FM is becoming busier every year and our core business and breadth of knowledge needed are increasing all the time.

Warning
Only monitor a service if it needs to be monitored

Ask yourself the questions: what benefit do I or the service get from monitoring something closely; does the amount of effort needed justify the outcomes; and is there another way of getting the information without needing to compile it?

- *What benefit do I or the service get?* Are you collecting information and watching a process for a reason? Can you quantify what benefit the service gains from you watching it? Would what you are monitoring do exactly the same whether you watched it or not? Do you have faith in the staff undertaking the process?
- *Does the amount of effort needed justify the outcomes?* Are you putting in a lot of effort into one area to obtain only a small or marginal improvement? Does the fact that you are monitoring something make it any more efficient? If not, then why monitor it? Could your time be best used elsewhere on an income-critical area that would give you a better return on your time?
- *Is there another way to get the information or manage something without your input?* Can you effectively delegate the task? Can you make any data collection implicit in contract documents so that the contractor pulls the information together or quality manages it for you?

Whatever you do however, make sure that your prime reason for doing it is to improve quality, not just to decrease your workload. Remember: "delegation, not abdication."

Quality management

Everything that you do as a Facilities Manager should be with a view to providing that service in the best possible way, building in quality as you go. This section aims to help you set up a QMS. Quality Management is not just a bolt-on service. It should be part of everything that you do within your department. Consider quality as just part of your specification for the work you do; it must be incorporated into everything and not just considered as an afterthought. What I mean by the word quality?

What is "quality"?

Quality in terms of the service that you provide is four things taken together: excellence, compliance, intelligence and review.

Excellence is the best that you can provide given the resources you have. Many people associate quality with the finer things in life and always they assume that it means the best that money can buy. This is not always the case. It is possible to provide an excellent service on a meagre budget; it is down to how you define excellence. Excellence is achieving the best possible outcome given the parameters set. So, if you have set the parameters of the service and you have driven the service to achieve them, then you have achieved excellence. We all know it is not as simple as that, because we rarely provide a service that cannot be compared against another provider. In the educational environment, excellence can readily be defined because we are compared through national statistics and surveys such as QDP.

Some cheat by setting the parameters of the service they provide too low, then by achieving those parameters claim they have achieved excellence. If you are not continually pushing the boundaries of your service by reviewing and increasing or improving your targets, you are not achieving excellence. Excellence is a self-driving aspiration because, for excellence to mean anything, continual improvement through development and self-analysis always move the goalposts. This is good.

Note to self

There is no excuse for a lack of quality in everything that you do. Just because your budgets are cut does not mean you can't still provide a quality service.

Compliance is more than just adherence to legislative requirements it is about adhering to your strategic objectives and driving quality forward by ensuring you set goals, plan how you are going to achieve your goals, action your goals and review your outcomes, which, in turn, will define new goals. This virtuous circle is the basis for most quality systems. Some people become blinkered and look on compliance as being simply legislative. Being narrow-minded in this field prevents you from achieving more than simple legislative compliance.

Compliance is also a form of audit. It is about driving quality through to the point where non-achievement is seen as non-compliance and is given the same significance as non-compliance to legislation.

Quality is not a blind, monotonous aspiration that relies on following a predefined path. It requires thoughtful, intelligent decisions to create outcomes. Systems work fine and can achieve very good results, but, unfortunately, as soon as you add humans into the equation your systems may fall down. Using intelligence is how a manager manipulates, cajoles and diplomatically pulls a diverse and disparate team together towards a common goal. Manipulates seems rather dictatorial; however, to get the best out of our staff we are forced into gentle manipulation to maximise their potential.

Intelligence is about adapting and keeping an open mind about what can and cannot be achieved with a group of people and setting your expectations to suit. For example, a solution or process that works for one set of employees may not work equally well for another set because of different characters within the groups. You may need to change the process to get the same result. Rigidly sticking to systems will reduce your ability to flex with the needs of individual groups.

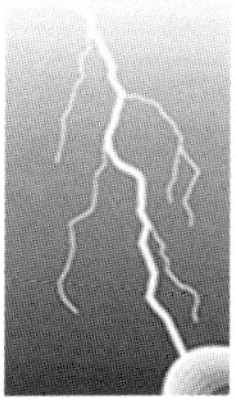

★★★★★★★★★★

Defining the level of service and therefore level of quality is essential to the success of your department. You should be striving for quality at every turn. Anything less is a missed opportunity.

★★★★★★★★★★

If don't review your outcomes, how do you know you have achieved what you set out to achieve? Review gives you the ability to define what you have done and should, if done correctly, allow you to see what you need to do next to start the whole process again. Review is, or should be, the springboard that allows you to take your service to the next level. Unfortunately, we often forget to review what we have done effectively, or give it just cursory credence and miss the opportunity to develop the service.

What is quality management?

Quality is therefore our driver and allows bench-marking and objective-setting. It is clear that quality should be used as a tool to continually develop your service. Quality is not a process with a beginning and a fixed end-point; it should allow critical analysis and prevent you from making the same mistake twice.

Quality Management is not always "high-end" service; it is about providing the best you can for the resources you have. If you provide a good service on a small budget, this proves you are working in a quality way. Throwing lots of money at a service should provide good outcomes, but are you getting the maximum return from your resources? Do not apply the word quality to a service that has more resources than it needs. You should substitute "quality" with the word "luxury" when applying that set criteria.

A quality service is:

- *Defined.* It has a set of parameters which are measurable.
- *Resourced.* It is funded sufficiently to ensure those parameters can be met.
- *Managed.* It is monitored and controlled to meet the measurable requirements.
- *Time-bound.* It is set out to hit key deliverables on time.
- *Reviewed.* It is evaluated to ensure targets have been met.
- *Improved.* Improvements are made to the service after reviewing it.

The key elements in setting up a quality system for FM are:

- *What is the service?* You need a specification for the service you want to provide. Does your service represent the strategic objectives of the organisation? If not, why not?
- *What elements of the service do you want to measure?* Also, what elements do you not need to measure? What elements of the service are critical? There is no point in measuring every output in a contract. Your system would be far too unwieldy. Do you really need any information other than service-critical measurements? Far better to pick a few key measurable parameters, especially those that demonstrate you are achieving strategic objectives.
- *What are your key deliverables?* What are your service-level expectations? Have you set the key deliverables for the contract?
- *How and when are you going to monitor and manage the contract?* This has been covered elsewhere in this chapter.
- *How are you going to review progress?* What meetings and evaluations are you proposing and are these clearly set out in a way that can be followed?

– *How are you going to verify progress?* Are you going to bring in an external person to verify that your QMS is being followed and adhered to?

Bench-marking

Bench-marking is the ability to draw a line in the sand, or to define a level of service. Bench-marking should be easy if your data capture has been good and consistent. It is a measurement tool and probably the most important tool if you intend to be able to prove the 'distance travelled'. Quality as a driver for any process needs targets and measurement. The ability to be able to accurately convert data into a definable percentage improvement is 'distance travelled' and the ability to measure your progress can convert plain data into tangible bench-marks.

You have to be able to measure apples against apples. Comparing data that is flawed in any way or missing vital pieces of information is bound to result in a difference in results, making your data useless. You need to be assured that

Theory and information slot

Used by many councils and local authorities, the "Golden Thread" is when an organisation has a common or shared purpose which is owned by all the employees "all pulling in the same direction." The organisational strategy is then aligned to this shared purpose to create and engender buy-in and ownership throughout every level of the organisation. Linking core aims and objectives and eliminating a behind closed doors philosophy builds a level of self-purpose and an engaged workforce.

Theory and information slot

Frederick Winslow Taylor (1856–1915) is regarded as the father of scientific management. He was the first person who felt that work deserved systematic observation. Taylorism had four principles:

> scientifically study tasks; scientifically select, train and develop employees; provide detailed work instruction; divide work practice equally between workers and management.

Taylorism, Buchannan, D.A. & Huczynski, A.A. (2010) Organizational Behaviour (7th edition) Pearson Education Ltd, Essex. (P 423) the information you collect this year is the same as the year before and the year before that. Without the assurance that your data are matched perfectly with previous data, your bench-marking information is flawed. At first, it may be that you have to settle for bench-marking as a principle and prepare your first year of data correctly as a baseline for the future.

However, this does not mean that you need to ignore previous data. There will still be certain aspects of this that you can reuse, such as meter readings for utilities and financial information on expenditure for maintenance or service contracts. There will always be valuable snippets of data within your department that you can take and turn into that notional "line in the sand."

Why is quality so important?

Quality is so important that if, as a manager, you strive for anything less than quality services then you are destined to fail from the start. Quality is important because if you provide a service that needs to be defined, you need to be able to assess whether you are doing it in the best possible way. How will this help you to provide a good service?

If you follow these simple processes and practices and can ultimately prove you are running an effective service, how can you ever be challenged about your work? If your service meanders aimlessly, not changing, not developing and not progressing, then you will probably find you will get fall into the abyss of outsourcing your service to a more proactive supplier.

If you are serious about quality and want, not only to be able to define your quality but effectively manage it, ISO 9001 is probably what you need to look at. ISO 9001 involves mapping out the key processes within the organisation, ensuring conformity and clarity of all the process undertaken within your FM department. This is a top-down look at your systems, starting from your strategic objectives and moving down through strategies, policies and procedures to guidance notes and forms. All sit comfortably within your QMS. Defining all your processes in this way creates clarity and uniformity across your provision in a way that is easily understood by your staff and which is clear and unambiguous to audit.

The principle of the auditable system is that it can provide you with clear bench-mark data on progress, allowing you to plot the distance travelled by your section. This is of great benefit in an organisation that is driven by the need to improve. Your system needs top-down commitment from management to succeed and a properly managed and audited system gives clarity of intention of the organisation to work in a quality way with an emphasis on continual improvement. To be successful, the system needs to be certificated and auditable by an external assessor. This is the key to any QMS.

Theory and information slot

Some of the benefits to your organisation of ISO 9001 are:

Provides senior management with an efficient management process

Creates areas of responsibility across the organisation

Mandatory if you want to tender for some public sector work

Communicates a positive message to staff and customers

Identifies and encourages more efficient and time-saving processes

Highlights deficiencies in some processes

Reduces your costs and time

Provides continuous assessment and improvement

Marketing opportunities

Some of the benefits to your customers of ISO 9001 are:

Improved quality and service

Delivery on time

Adoption of the getting it right first time attitude

Fewer returned products or failed services and complaints

Independent audit demonstrates commitment to quality

Risks and liability

Compliance is a huge part of the Facilities Managers' role. We need to make the site safe and secure, not just for staff, but also visitors. This is a duty that can easily be hidden, only coming to light when an inspector arrives, by which time it is usually too late. Better to be ahead of the game and have a system in place which helps you to monitor compliance. This can be linked to your QMS.

What is compliance?

Compliance is a positive response to legislative requirements, i.e. actually undertaking what individual items of legislation require and monitoring and managing that process. It is simple to say that legislation is easy to comply with; however, legislation rarely explains everything you need to do. It is generally sprinkled liberally with the phrase 'as far as is reasonably practicable' and while this phrase does give some latitude, this can lead to misinterpretation.

Your policies and procedures should provide a framework to meet compliance obligations, so when you review your procedures it is essential to try and turn "interpretation" into "prescription." This will allow you to define clearly what you need to do to comply with the legislation. It will also provide an audit guide. Interpretation is never easy to monitor and make judgements on the quality of work.

Statutory requirements

By far the biggest risks from the perspective of your organisation are the legislative requirements. These will see you in court if they are not met. The big three of the building requirements are: asbestos, *Legionella* and electrical testing.

Asbestos

There are still a number of buildings which contain asbestos in some form. This is far more prevalent in educational establishments because of the post-war boom in school building around the time when asbestos materials were seen as the most economical to use for insulation and fire protection; this solved one problem and created another.

From your perspective, the essential things you need to know are:

- where are the materials containing asbestos?
- what type of asbestos is the material?
- what condition is the asbestos in?

Aide Memoire: policies, procedures, risk and liability

- Review all policies and procedures to ensure there is completeness and that, from policies through strategy to forms, make sure that there is a logical flow.
- Look at all of the risks and liabilities of the organisation and understand how you interface with these risks and liabilities and how you can affect them.
- Establish how you will monitor and manage your service and set up some key dates for your monitoring and managing tasks

Establish your levels of quality and work out how you will monitor and ultimately improve quality.

- how do you prevent release of asbestos fibres?
- how do you monitor degradation?

The only way you can do this is by the following four-step process.

Survey

Commission a survey to identify the location something you can do yourself and you should not even try unless you have had the required training.

There are many materials similar to asbestos and the only way to confirm its presence is for samples to be taken under controlled conditions and then tested in a laboratory. Only then will you have the definitive answer to what kind of material you have on site. There are many companies who undertake these surveys. Make sure that they are certified by one of the accredited organisations such as the National Asbestos Training and Accreditation Scheme Ltd (NATAS). It does no harm to obtain competitive quotations for this work as it can vary widely in price.

Your survey will indicate where all of your asbestos-containing materials (ACMs) are and will also give you details of what condition they are in, giving you, in risk assessment format, a matrix on whether they should be removed or managed in situ. Wherever possible, seek the solution of managing them in situ. It is far more economical than full removal.

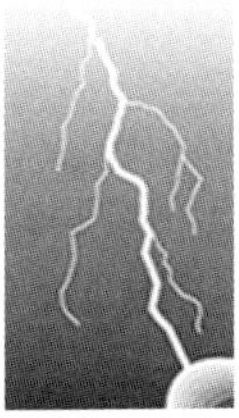

★★★★★★★★★★

Make sure when you tender for the identification of ACMs that you make it clear you will be tendering their removal separately. Contractors will then have no incentive for finding more ACMs.

★★★★★★★★★★

Removal

Sometimes you may be forced into removal and, if you are, these should be tendered works to make sure that you get the best price. Your contractor will be able to advise on how long the works are likely to take, what impact they will have on the buildings and whether you will be able to continue to occupy them during removal. Be careful to factor these items into your considerations. You do not want a building out of operation at business-critical times of the year. You should only remove items that are considered dangerous if left. Do not allow people to persuade you into accepting that all asbestos is too dangerous to be left in situ. It is only dangerous if it is naturally degrading, i.e. flaking, or if it is in an area where constant contact by people is inevitable, or constant maintenance operations make it likely that it would be disturbed, allowing asbestos fibres to be released into the atmosphere. If you are in a position to manage and control access and the condition of the asbestos, then this is a far more economical way of looking after ACMs.

Management

Management of ACMs may seem onerous, but in real terms is relatively simple. You need to know where the materials are and this is the purpose of the initial survey. You then need to exclude activity from the area of the ACMs or place a barrier between the user of the area and the ACM. With asbestos sheeting this could be as simple as painting the asbestos board to seal it from flaking or releasing fibres into the atmosphere. It is also a sensible precaution to mark the board as containing asbestos using the official stickers.

You must also have a management system in place to prevent maintenance contractors from disturbing the ACMs. This system should allow contractors visiting your site to have access to the asbestos register. If you are aware of asbestos in a building and you do not declare this to the maintenance contractors, then you will be liable for any injuries they sustain, even if the contamination takes many years to manifest. It is better to be safe than sorry and to disclose ACMs to every contractor and with any tender that may include potential disturbance of the ACM.

If your staff are likely to come into contact with the ACMs, then they will need to have tests to determine whether they have been contaminated and then to assess the degree of contamination. This is essential in determining if your workforce has come into contact with ACMs before their employment at your institution. These tests must be undertaken when first appointing them to the position. You also need to arrange staff training to ensure that they are aware of the dangers of ACMs.

Maintenance

You cannot leave a dangerous substance like asbestos in a building without a long-term maintenance strategy to ensure that it is not disturbed and that it does not degrade and become dangerous. You need a repeat inspection of the ACMs to determine their condition. There is no fixed detail on the frequency of these surveys, but I would recommend every three years. These should be undertaken by an external organisation. The advantages of using an external organisation are:

- You can forget about it to some degree, as your survey provider will advise you when your next survey is required, allowing you to concentrate on other issues.
- They will have more experience than you in undertaking these surveys.
- From an audit perspective, it is more credible to have an external agency verify your survey results.

Despite the fact that you may buy in expert help to undertake repeat inspections, you still have an onerous duty to ensure you monitor the ACMs. Part of this duty is to periodically check if items such as stickers are still in place, or if obvious damage to areas has taken place. It is prudent to record these minor inspections.

Your management system must look at how you prevent the ACMs from further damage. This is primarily about controlling access to these areas by contractors and staff. Having a full set of survey results available gives you the knowledge of where everything is. It is then prudent to keep an access record to these areas indicating when contractors and staff have worked on or near the ACMs.

It is also essential that you keep a record of any breaches in your system in terms of accidental disturbance. This should contain information such as who disturbed the ACM, when and to what degree it was disturbed and what action you took to remedy the risk. However, if an area of asbestos is that easy to disturb, you should seriously consider having it removed. It will be far less costly than dealing with litigation from someone who has been contaminated with the ACM.

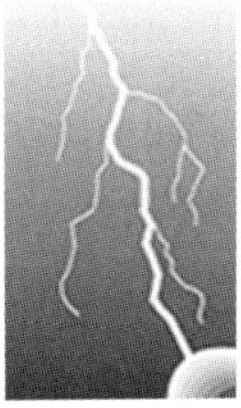

★★★★★★★★★★

As with asbestos, when you tender the risk assessment survey for *Legionella*, make it clear that you intend to place any remediation works with a different contractor.

★★★★★★★★★★

Legionella

Legionella is one of those problems that sits quietly in the background and occasionally comes to the fore when a major outbreak brings it back into the headlines. *Legionella* is usually dealt with under PPM. Managing *Legionella* is about controlling the environment where the bacteria can multiply and become a risk to your site. There are basically two areas that you need to concentrate on in terms of being compliant with legislation: survey and identification and management. I am not going into detail about what *Legionella* is and the dangers associated with it. My purpose here is to provide an overview of the problem.

Survey

In the same way as with asbestos, you need to know where the potential sites for *Legionella* are within your building stock. The best way to achieve this is to obtain a survey of the entire site by an external contractor to determine:

- the condition of the cold water tanks
- the location of the sentinel taps on the system
- schematic diagrams and layouts of the system
- identification of any dead-legs in the system
- the location of the shower heads

This then becomes your risk assessment document for the site, laying out the possible risks for the site in general. Surveys can be relatively cost-effective and take away the guesswork and can also provide a useful reference guide for your plumber.

Make sure that you write into your terms and conditions that the incumbent survey company provides awareness training to your staff as part of the package. Most *Legionella* management companies will be pleased to provide the training free of charge.

Management

This is one area that you need to be watertight on (pardon the pun). Your management will fall into the following areas, all of which can be covered by simple PPMs: sentinel taps, blowing down calorifiers, running of infrequently used taps, removal of dead-legs and shower head disinfection.

- *Sentinel taps.* Temperature tests need to be performed on your first and last taps on any system; these taps are known as sentinel taps. These temperature readings must be logged and records maintained.

Inform the Facilities Manager of any water outlets (taps, showers, etc.) that are not used. These can either be disconnected or removed. Any equipment that stores water should be cleaned, drained and replaced in accordance with the manufacturer's instructions and infection control protocols. Ensure everyone seeks the advice of the Facilities Manager before purchasing any equipment that uses water.

Note to self

Start to pull together information from the O & M files. If you have new buildings the original commissioning certificates for the electrics will be in there and you don't need to electrically test these buildings until they are five years old. Be careful in relation to landlord's responsibilities as these are different.

- *Blowing down calorifiers.* This needs to be undertaken on an annual basis and will ensure that any build-up of sludge in the bottom of the calorifiers is removed.
- *Running off infrequently used taps.* Outlets that are not used at least once a week and cannot be removed should have all sources of water run off for at least three minutes, at least on a fortnightly basis this action and documented on a record sheet. This includes outlets that have been taken out of use on a temporary basis.
- *Removal of dead-legs.* Redundant or very infrequently used outlets should be removed from the system. You will often see sinks that have been removed from rooms without fully taking out the redundant pipes back to the nearest in-use line. This is often done because "we could need it later" or "they will ask for it to be reinstalled within a year." However, these sometimes long dead-legs can become the breeding ground for *Legionella*. It is better to disconnect the outlet and leave the pipes in situ. At least the pipes won't have water in them.

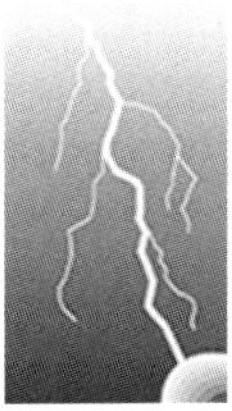

★★★★★★★★★★

Make it clear in your tender documents that the testing and any remedial works will be dealt with by different contractors to remove the incentive to the contractor to find issues and faults.

★★★★★★★★★★

- *Shower head disinfection and cleansing.* Shower heads often sit unused for long periods, particularly if your institution does not operate during the holiday periods. There is a potential for *Legionella* bacteria to build up in shower pipes and heads, so flush them out, remove the heads and disinfect them.

Electrical testing

Owners and landlords of buildings have a duty to ensure the electrical safety of the fixed wiring of a building. By fixed wiring, this means everything up to the plug of an appliance resting on your ring main. All non-domestic electrical installations are subject to a regular programme of testing and inspection to comply with current legislation. You must ensure that all necessary precautions are taken to reduce the risk of injury or death through electrical failure. This is a requirement under the IEE Wiring Regulations (Guidance Note 3: Inspection and Testing, 6th edition, 2011).

Sometimes known as periodic testing, these tests should be carried out on your installations on a cyclical basis. This may also be referred to within your fire risk assessment, so it is one that you really must comply with. Try to structure your tests over a period of time rather than a single one test period; this reduces your budgetary impact and allows you to smooth your budgets over a period rather than introducing peaks and troughs. You are obliged to keep these records for viewing if required and this will probably be one of the first documents your insurers will ask for if you ever have an electrical fire on site. Refer to the IEE Guidance Note 3: Inspection and Testing, 6th edition, 2011 for details on periodicity, as this varies depending on the type of installation.

> **Note to self**
>
> You must keep accurate documents for all our compliance issues.

There are two types of test. A routine check ranges in frequency from every four months to every year. A full inspection and test ranges in frequency from at each change of occupancy for a domestic property, to every six years for highway power supplies, but more generally every one to five years depending on the type of installation.

Other areas for health and safety are detailed in the following table.

Risks with new builds and defect periods

Your new build is a perfect example of building technology. It fits your users perfectly, has no defects and saves on your maintenance bill for the next five to ten years. If you believe that, then you are working under a big misapprehension. Your building could provide you with a nightmare from the day it first opens.

Areas of risk

Area of risk	Relevant legislation	Reference
Slips, trips and falls	Workplace (Health Safety and Welfare) Regulations 1992	HSE (2005c) Preventing Slips and Trips at Work. Series code: INDG225 (rev2) (2012) www.hse.gov.uk/slips
Hazardous substances	Control of Substances Hazardous to Health Regulations 2002 (COSHH) amended	HSE (2009b) HSE (2007a) www.coshhessentials.org.uk
Fire	The Regulatory Reform (Fire Safety) Order 2005 The Dangerous Substances and Explosive Atmospheres Regulations 2002 (DSEAR)	Controlling Fire and Explosion Risks in the Workplace. Series code: INDG370 (rev1) (2013)
Falls from height	Work at Height Regulations 2005 Provision and Use of Work Equipment Regulations 1998	HSE (2007b) HSE (2008c)
Lone working	Health and Safety at Work Act 1974; Management of Health and Safety at Work Regulations 1999	HSE (2009a)
Musculoskeletal disorders	Manual Handling Operations Regulations 1992 (as amended); Management of Health and Safety at Work Regulations 1999	HSE (2003b) HSE (2002) HSE *Manual Handling at Work.* Series code: INDG143(rev3) (2012)
Noise	Control of Noise at Work Regulations 2005	HSE (2005a) HSE (2005b) www.hse.gov.uk/noise
Work equipment and machinery	Provision and Use of Work Equipment Regulations 1998; Lifting Operations and Lifting Equipment Regulations 1998; Supply of Machinery (Safety) Regulations 1992 (as amended) 1994	HSE (1999) HSE (2008b)

(*Continued*)

Areas of risk (*Continued*)

Area of risk	Relevant legislation	Reference
Workplace transport	Workplace Health, Safety and Welfare Regulations 1992; Provision and Use of Work Equipment Regulations 1998; Lifting Operations and Lifting Equipment Regulations 1998	HSE (2005d) HSE (2000b) HSE (2005e) www.hse.gov.uk/workplacetransport
Asbestos	Control of Asbestos Regulations 2012	HSE (2006a) HSE (2006b) www.hse.gov.uk/asbestos
Legionella	The Health and Safety at Work etc Act 1974; Control of Substances Hazardous to Health Regulations 1999	HSE (2000a) www.hse.gov.uk/legionnaires
Electrical safety	Electricity at Work Regulations 1989; Provision and Use of Work Equipment Regulations 1998; Supply of Machinery (Safety) Regulations 1992; Electrical Equipment Safety Regulations 1994; BS7617 2008–11 Requirements for Electrical Installations IET Wiring Regulations 17th edition. BS7671: 2008 (2011).	HSE Safety in Electrical Testing at Work. Series code: INDG354 (2002) HSE (1998) HSE (2003a) www.hse.gov.uk/electricity
Gas systems	Gas Safety (Installation and Use) Regulations 1998; Approved Code of Practice and Guidance	HSE (2011b)
Local exhaust ventilation	Control of Substances Hazardous to Health	HSE (2011a) HSE (2008a) HSE (2008d)

User requirements

You will have undoubtedly gone to great lengths to ensure that all departments have had an opportunity to be part of the design process. They will have helped you to develop the requirements of the department or function. They will have signed off room data sheets and had the opportunity to review their designs, and will have been part of the process from start to finish.

You cannot avoid the precept that "I didn't think it would look like that" or "it's difficult to see what this would look like on a flat plan." The truth is that no matter what you have done, what suited the department 18 months ago when the process started may not suit the department once the building is complete. You must not be disheartened to learn that you could be undertaking adaptations to your new building soon after the opening day. This is not being critical of individual departments for being indecisive; it is inevitable that change will affect your best laid plans.

Users expect that your bright new building will be the answer to their prayers. They will be far less tolerant of high or low temperatures or a lack of, or too much, ventilation. Your commissioning of the final heating and ventilation service must be very carefully undertaken and any anomalies dealt with as early as possible while you still have the bargaining tool of the defects period.

Heating systems

New and modern systems are far more likely to give you problems than the more traditional system. When accepting a new building, you need to concentrate on training. Your new building will undoubtedly be filled with new technologies and your heating system will be at the hub of these, whether it is a traditional boiler, a ground source heat pump, photovoltaic panels or solar thermal. You will be required to ensure that the system is maintained and run effectively from day one. You could outsource this requirement to an external company; however, this is costly. It may be better to have your own staff looking after the equipment.

> **Note to self**
>
> Make sure that you obtain the full commissioning certificates from the construction company and ask them to prove the validity of the commissioning figures.

> **Note to self**
>
> Never expect your shiny new building to have zero defects and don't be lulled into thinking you won't need to make adaptations in year 1.

This gives you the control and responsiveness you will require.

However, you cannot expect your staff to be able to deal with this without a thorough understanding of the system and how it is operated. To get the best out of the system, try to arrange training from day one, with a refresher course later in the year. If possible, build this into your new build tender documents to ensure you don't end up paying for it as a separate item. Try to make sure the training spans two seasons, i.e. winter and summer, as you will find your system operates differently at different external temperatures and it is far easier to resolve training issues when faced with the greatest range of operating conditions.

Make sure you are building in PPM from day one and make sure your PPM is split into that undertaken by external contractors, such as boiler maintenance, and that undertaken by in-house staff, such as blowing down calorifiers. This way you are less likely to miss the essential items needed to use your new systems effectively. The Appendices to this book give a strategy for heating and ventilation.

Fire legislation

You will probably find that within a few days of opening your new building, various objects will appear in corridors and hung on walls that do not comply with the initial fire strategy for the building. It is imperative that this is dealt with decisively in the early days before it gets out of control. Inevitably your Fire Safety Officer will meet with you early in the operational life of the building. If they find you have ignored the fire strategy and filled all the fire-sterile areas with combustible items, he or she is not going to be very happy with you. Long-term relationships with your Fire Safety Officer can be made or

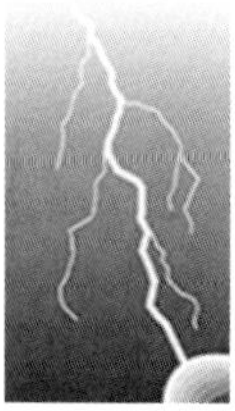

★★★★★★★★★★

Make sure your construction contractor gives your on site staff full training in the use of the systems within the building; be cheeky and ask for a refresher after six months.

★★★★★★★★★★

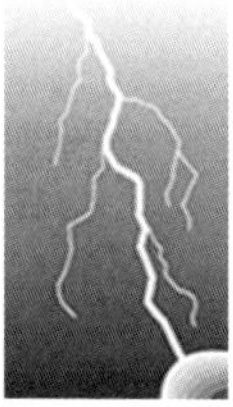

★★★★★★★★★★

This document is best pulled together by an unbiased external company. This gives you credence and validates/audits your fire precautions. Many ex-fire officers have formed companies to undertake this service and, in my opinion, these are the best to undertake the work,

★★★★★★★★★★

broken by your actions on issues like this. If you show a decisive attitude you will find they will leave you alone; if not, you will experience an increased frequency in their visits and far more scrutiny.

Fire risk assessment

A key document in compliance, your fire risk assessment, if done correctly, can meet legislative requirements and make your life easier. Your fire risk assessment is an overview of fire safety arrangements for your building and also details any potential pitfalls with construction and subsequent works. It is a live document, which means it is never really complete. You need to add to it all the time.

Key elements of your fire risk assessment are:

- How you deal with fire as an institution and what procedures you have in place.
- Who is responsible for what elements of the document.
- What dangers exist and how are they dealt with.
- What defects exist and how and when you will deal with them.

Although the fire risk assessment is a live document, there are aspects of that document that are snapshots at the time of inspection. They will identify flaws with your processes, procedures and practices and look at the physical aspects of the building that may have changed. Therefore we can logically split these into two distinct areas: fixed detail and live detail.

Fixed detail items are those which, under normal circumstances, will not change, such as procedures, contact names and responsible persons. These are generally the basic details of compliance in terms of how you intend to deal with fires and what standards you set internally to monitor yourselves against. Live detail items are the results of inspections and tests, e.g. the fire log book, which details when items have been tested and by whom. Included in this list are:

- Fire alarm tests. In an educational environment these should be on a weekly basis rotating call points, and also include sounder checks. This gets the users of the building used to the sound that the fire alarm makes, so that in the event of a fire so they do not panic.
- Tests on fire curtains and smoke removal fans.
- Details of external contractors, such as service and maintenance details.

- Annual training of staff and induction.
- Physical details of faults found.

Faults found

Your fire risk assessment is generally an annual process and your nominated risk assessor is generally bought in to undertake this. Do not feel aggrieved if, at your re-inspection, you have a long list of non-compliance items. Fifty-two weeks is a long time in any business. Things can happen to your site that should be under your control, but they happen because you cannot stand over people 24/7. The types of faults you will generally find are:

- *Poor use.* People propping fire doors open with door wedges, or obscuring fire safety notices.
- *Failing items.* For example, door closers not closing properly, smoke seals fraying or failing out.
- *Non-adherence to procedures.* Missed inspections and tests.

Theory and information slot

Five steps to risk assessment

The established Health and Safety Executive five-step process is the industry standard.

- Identify the hazards.
- Decide who might be harmed and how.
- Evaluate the risks and decide on precaution.
- Record your findings and implement them.
- Review your assessment and update if necessary.

A hazard is anything that may cause harm, such as chemicals, electricity, working from ladders or an open drawer and the risk is the chance, high or low, that somebody could be harmed by these and other hazards, together with an indication of how serious the harm could be.

The five-step process can be found at http://www.hse.gov.uk/risk/fivesteps.htm

- *Faults to structure caused by others.* For example, Intumescent strips being painted over by your painters or your IT contractors knocking holes in walls and then not using Intumescent foam to re-establish the fire integrity of the structure.

The risk assessment process is really a subject in itself and is only touched on this book. This key task for the Facilities Manager needs to be addressed fully as most compliance issues are dependent on the risk assessment process.

Emergency procedures

Any organisation needs to be sure that, in the event of emergency, it is able to deal with that emergency effectively. This is a requirement incumbent on all organisations. There are a number of situations that would require a response and the main ones are:

- fire evacuation
- intruders
- bomb threats
- business continuity

All these examples can be dealt with in the same way. Looking at each one as a risk assessment is the simplest and most effective way of dealing with the issues. The key elements to be addressed are:

- identify the hazards
- decide who might be harmed and how
- evaluate the risks and decide on precautions
- record your findings and implement them
- review your assessment and update if necessary

All emergency processes must be fully communicated to staff to be effective.

Risks to property

Most of the risks you will have to deal with will centre on your property. Your building stock and the quality of that stock is your responsibility. It does not repair itself and it does not self-identify problems. This is your responsibility and yours alone, so if things go wrong the problem sits squarely on your shoulders. That is not to say that things can go wrong which are outside your control, but in general you should know about these in advance. In terms of risk aversion, there are two main areas to be considered: site building maintenance and planned or capital risks.

Building maintenance

> **Note to self**
>
> A desktop survey can pull together a huge amount of the data you will need for your planned survey if you have good records.

Building maintenance is often overlooked when considering risk aversion, but it can be one of the areas, if treated correctly, that can significantly reduce your risks and liabilities on site. Prompt repair and maintenance helps to reduce the potential for breakdowns and prevents a fault becoming worse. There are basically two types of maintenance when considering risk aversion: reactive maintenance and PPM.

Reactive maintenance is responding to repairs as they occur and is unavoidable, as things will always break. It is how you respond to those breakages and in what time frame which increases your potential for risk aversion. Prompt repair is essential and shows a professional attitude from your department. However, getting a repair done can be a paper chase, with requests for repairs being passed from pillar to post, inspected and re-inspected before the work is undertaken. You must try to reduce repair times to a minimum by removing these layers of administration and concentrate on getting the job done quickly.

Reactive maintenance can be a massive drain on your budgets and will, over time, increase as your building stock ages. If caught and dealt with early, these drains on your budget can be minimised.

Planned preventative maintenance is the most important type of maintenance to master and all too often it is missed or forgotten and ends up being reactive maintenance. A typical example is when filters on air-handling units clog up and this then flags up a failure on the BMS. The filters are replaced and the repairer assumes that he or she has undertaken PPM. All too often this is the norm, not the exception. PPM can save money in the long term, but, as a department, you need to deal with it properly, not pretend that you are managing it. This is dealt with further in Chapter 7.

Planned maintenance

Your approach to planned maintenance is one area where your risk aversion can be plotted in advance. Determining where the risks are for the future (and you must include the risk of needing funding to deal with future problems; risk is not simply health and safety) is the key to being proactive about the replacement status of your site.

The collection of data to confirm and validate your conclusions is the key to staring this task. This process is huge and could be a topic for its own book. Here, I am dealing solely with the risk aversion elements of planned maintenance. Your process of data collection should be as follows: data collection; planned survey; analysis of data and survey; planned programme, including property risk register; action; and review.

Data collection

To get the most out of your survey, you need to have at your disposal all the essential data that could influence your decisions as you progress. These may include items such as:

- *Plans of the building.* These are a useful way of making judgements about the structural elements of the building and allow you to see the interconnection between various elements. Plans are great to use as a sketch pad as you carry out your survey. Using a copy plan allows you to annotate problems as you proceed; this then becomes an aid when you get back to your desk. Plans can also help you to see services and M & E in a much clearer way than just simply lifting a few ceiling tiles here and there.
- *Maintenance records.* Sometimes maintenance records can give a great insight into the condition of plant and can give you warning of equipment coming to the end of its life span, e.g. a greater frequency of breakdown in a lift could indicate that major refurbishment is required.
- *O & M manuals.* These are one of your most valuable pieces of information relating to the building. They should provide all the data you are likely to need, if they have been set up correctly, such as: manufacturer's warrantees; maintenance regimes, as sometimes critical maintenance operations or major replacements can be listed in the maintenance regimes section, e.g. the replacement of glycol in

your ground-source heat pump between years 5 and 7, which would be a major unplanned cost if you didn't factor it into your planned programme; and manufacturer's anticipated lifespan for equipment. Manufacturers of M & E equipment in particular will suggest a lifespan for that kit, e.g. boiler plant could have a 20 year life span. It is clear you need to plan well in advance for items such as this.

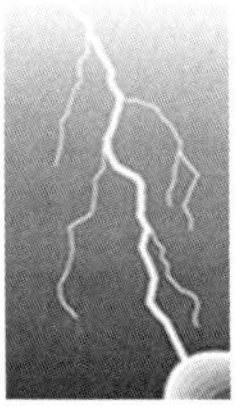

★★★★★★★★★★

Get someone else to do all the hard work; use your service providers to give you details of the life expectancy of the various elements they maintain.

★★★★★★★★★★

Once you have all of your information in one place it will be then possible to get ready to verify some of your data and check the condition of the various elements. Pulling all the elements into one coherent document is really what you are trying to achieve. Your document should be in two parts: data collection and executive summary. It is easiest to use an Excel spreadsheet for the data collection element of the work. This can be organised into the constituent parts in a logical way; breaking things down into bite-sized chunks makes it easier for you to pull the information together and then helps to justify or verify your conclusions by giving your reasons in an easily understood format. Your survey can now begin in earnest and will be in two parts: a desktop survey and a physical survey.

Your desktop survey brings all your paper evidence, such as O & M records, together and starts to break down the individual elements of the building into those bite-sized chunks. However, the physical survey is you getting out there and getting dirty. This could involve going up onto the roofs (please remember safety; conduct your risk assessment beforehand and ensure you have the correct access equipment) or, better still, getting somebody else to do it for you. You will find that your local roofer will only be too pleased to give you the benefit of their advice and do a full survey for you, sometimes for free, particularly if you promise to put them on the tender list when the works are ratified. You can use this principle for all of your major building elements.

You should have in place the essentials such as boiler and ventilation maintenance contracts; use these to give you an appreciation of your equipment or build this into your tender documents when you re-tender the service for an annual appraisal. If you settle on a format for the return of information to you, then you have a consistent reporting system with which to update your planned programme annually.

Analysis of data

All these data need to be pulled together in a format that is easy to understand and manipulate and easy to add to or delete from. At this point you sometimes need to add in an element of subjectivity and reality. When you start to pull together your data, you will start to come to some conclusions about the cost of replacement and when items will need to be replaced. What you need is a smooth flow of work over your prescribed period; you don't want all of your replacements bunched into one year, nor do you want a huge financial commitment in any one year. Your skill as a Facilities Manager should be to move items around so that you can smooth your resources over a period of time.

Planned programme

When you have finished allocating years and costs to your various projects you will have, the planned programme of works needed to bring your building stock back up to an acceptable standard. In an education environment this is known as Condition B.

Your planned programme needs to be understandable and easily digested. There is no point using esoteric language or acronyms; people need to be able to understand your document. You need to be able to appeal to the Finance Department more than any other, so make sure your calculations are correct and that your statements are simple, yet factual, otherwise your report will lose credibility.

Action

Your next action is to submit a lucid and easy-to-read executive summary for general consumption. If you present your full survey to your executive, you run the risk of diluting the intentions of the report, which are to identify those elements which pose the most risk in terms of health and safety and finance. A five-year summary focuses your efforts into identifying the key elements that are going to fail in the next five years.

Five years is about the furthest into the future you can predict with any certainty. You may wish to project an analysis further forward if you want to demonstrate a buildings obsolescence point, but, for the major elements within a building,

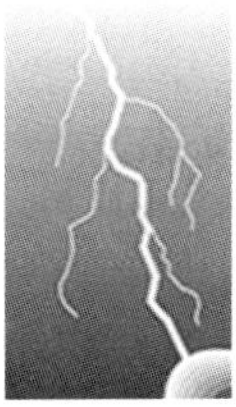

★★★★★★★★★★

Keep your planned data in bite-sized chunks. This keeps it far more manageable and easier to manipulate. It also makes it easier to alter at a later date.

★★★★★★★★★★

five years is the limit of accurate predictability. Your document should be live and able to be updated an annual basis. In subsequent years it will only take a nominal amount of work to update.

Your finished document needs to be dynamic and easy to read. It should be colourful and descriptive; boring documents are very rarely read, so make it as vibrant and interesting as possible. Remember, this is your chance to shine. If your target audience actually reads it you are more likely to get a positive result out at the end of the process.

Remember to undertake a cost-in-use exercise on the outcomes of replacing certain pieces of equipment. If you are not careful, you will miss the essential payback period.

Payback periods

> **Note to self**
>
> Don't expect your staff to stay compliant without you detailing their individual training requirements.

This is the money you can save by replacing something with a more efficient piece of equipment. This could be more energy-efficient equipment which, over a period of time, uses less electricity. Or it could be something that requires less maintenance over a period of time. All these are essential reading for your Head of Finance, who will love items that pay for themselves. Never forget, however, that failing items cost more to maintain and the cost of premature failure to the organisation also needs be factored in, such as the cost to the organisation of a major boiler failure resulting in the closure of a building in the middle of winter. All these should feature on your risk register and they also give more credence on your report for long-term strategies. You should aim for every element of your replacement schedule to and pay for itself, although this is not always possible.

Review

Your document should contain a review element to check at set points that the data you have are correct and any updates have been included. There is no point in spending time and money in undertaking a full planned survey if it sits and gathers dust in a corner. Use it wisely and it will secure the long-term future of your stock and will make it easier for you to pull together information quickly should your executive team want an update on progress. I t also clearly defines you as a proactive Facilities Manager, not one who simply eats into budgets by dealing with everything in a knee-jerk fashion.

Staff

Your staff can make or break you and are also your greatest liability in terms of their actions or inactions. They are capable of wrecking any process unless you are able to harness them to the maximum. Your staff need to be managed; this is an area you cannot delegate or abdicate from. You need to nurture and control them, while at the same time allowing them to grow and contribute to your department. In terms of risk and liability, you need to concentrate on three areas of staff management: training, certifications and monitoring.

Staff training

You cannot expect your staff to keep up with compliancy issues without devoting time to training. Training is not just a methodology for teaching someone how to do a job. It is primarily about training staff to understand the risks and liabilities of the functions that they undertake. This may sound cynical, but your staff will never be compliant without understanding the effects of their actions or inactions. Only by understanding the personal risk to themselves will they be motivated to learn and make sure they understand the implications of their actions.

Your training therefore needs to be structured to cover all aspects of their positions. This can be broken into two distinct areas: organisation compliance and personal compliance.

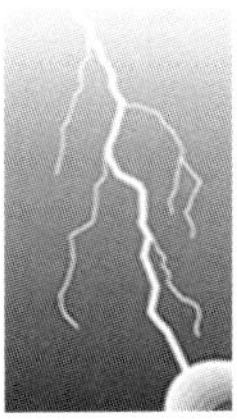

Organisation compliance

As an organisation, you have certain obligations in relation to your workforce and this needs to be conveyed to your staff in terms of formal, auditable training. These training sessions are for subjects such as: COSHH, from the perspective of personal contamination; manual handling (protection of your staff from musculoskeletal injuries); and general health and safety. This last area is sometimes forgotten. Your workforce needs to understand what health and safety actually is, otherwise they will not understand how the various elements fit together and interrelate.

Put together a list of all functions within your department, allocate training to each discipline, and then finally allocate staff to each element. This way you can see your training matrix holistically and make sure all are trained to ensure compliance is maintained.

Personal compliance

At this level, you are moving some of the responsibility of an organisation onto the workforce. This may sound cynical, but your job is to protect the organisation as well as your workforce.

Listing your training

It is worthwhile compiling a full list of the training your staff will need (refer back to the training matrix in Chapter 3). . For this purpose it is best to list out all the jobs undertaken on site and then decide what training, and to what level, is required. Only then can you allocate names to this list. This is perhaps best done as an Excel spreadsheet, with a list of tasks across the top and the training necessary underneath, then a list of all your staff down one side. It is then simple to allocate staff to both tasks and training. This can then form part of your appraisal process and will save you having to think about development tasks for staff, allowing you to concentrate on their personal development.

Certifiations

Overlay your training matrix with details of any certifications your staff may need. Certain members of your staff, particularly those in particular specialist areas, may need to update their qualifications on a regular basis, e.g. Gas Safe or NICEIEC. Don't forget to put these onto your training matrix. Even though they may only come up every two or three years, you need to factor them into your budgets.

Monitoring

It is not enough to simply assume that, because your staff are fully trained, they will act in your, or even their own, interests all the time. You need to monitor their personal health and safety. It must be made clear to your supervisors that this is exactly what they are there to do! Sometimes, when all is running well, your supervisory staff will take their eye off the ball and let things slip. Don't forget you are there to monitor and supervise them as well.

Never underestimate training as your route to compliance; in fact, you must make it your cornerstone. Without training, your staff are ticking time-bombs and a lack of training will come back to haunt you. Therefore devote time to it and, once in place, it will allow you time to concentrate on other issues and may just stop a few sleepless nights.

Financial risks

> **Note to self**
>
> Stay legal and ethical at all times; don't bend financial rules and make sure that all your staff follow the same principles.

Everything you do within your department is bound by finance. You cannot operate without available and accessible budgets. Therefore when you have the finance to undertake your work, you need to show due diligence for the work you do and the money you spend. Some of you will have large and complicated budgets and you need to be able to control these effectively and show an audit trail for everything that you do. This may sound obvious, but how many times in a busy department do you have to just do things and think about the finance later? This can leave you vulnerable if it becomes the norm rather than the exception.

Financial management

How you manage your budgets may ultimately be dictated to you by the financial procedures of your organisation. Compliance from the perspective of procedural management of your budgets will then be far easier. Even your tender and quotations procedures may be dictated by procedural management. If this is the case, then your life should be easier. However, there is usually a series of grey areas where procedures end and interpretation begins. It is these areas where you can easily fall foul of a lack of regimented adherence to due diligence. For example, there are bound to be certain aspects of the overall procurement process where you are able to influence the process. Unless you set yourself some rigid guidelines you can very easily find yourself in trouble.

Some things to avoid:

- *Avoid tandem systems to your financial system.* There is always the temptation to believe that you can control your finances better than your finance department. Invariably, your formal financial systems do not work in the present and can sometimes be as much as a month out of date in terms of predicting your current position. This has led to Facilities Managers becoming frustrated because accurately predicting your outcomes when you have large budgets is critical. Most modern financial systems provide accurate predictive outcomes. Talk to your Finance Department and tell them the kind of information you need. There is every chance that they will be able to give this to you as they can manipulate data to give you predictions of outcomes.

- *Not knowing the extents of your commitments.* When agreeing contracts, make sure that your specifications are correct. If you commit yourself to somcthing without tight specifications, you might as well sign a blank cheque.
- *Never ever accept gifts.* Even a bottle at Christmas can be a step too far. You cannot be impartial with a contractor or supplier if you have accepted gifts. If you accept a bottle at Christmas, declare this to your Finance Department. Even then, don't be tempted to take it home yourself. One trick is to pool all the bottles and have a draw with all your staff. You may even win a bottle, but at least you will have been fair. However, never assume that even this is acceptable and check with your Finance Department to make sure you are clear and clean. If you never accept this type of offer, you can never be challenged on it. Sometimes contractors working as part of your team invite you out for "team-building events or meals." Again, avoid these as these they could be construed as conflicting a interest, and how could you prove otherwise?

> **Note to self**
>
> Avoid rollover contracts that go on forever. You lose control of the specification and control over expenditure in the long run.

- *Closed tenders and quotations.* Always operate a system which guarantees the anonymity of the tenderers. Make sure your tenders and quotations come in at the same time in unmarked envelopes and that they are signed in by reception staff. Be rigid about late tenders; don't allow them. It will teach the contractor a lesson and they won't do it again. You cannot be challenged that you have given preferential treatment to one particular contractor. Your Finance Department will have some specific rules for each organisation. They may all differ slightly, but will follow the same principles of financial probity.
- *Avoid endless re-negotiations on the same contract.* A contract will often run for years because it is easier to say to the contractor "just give me an updated price." If you do this, you cannot prove that you got the best price for the job. You could be accused of favouritism and, just as importantly, your specification may have drifted, e.g. you may have added or removed buildings, or have not accounted for updated legislation or changes in practices. All of a sudden you have a contract and you don't know what services you are paying for. Remember you are in control and you will be accountable.

Fraud

Fraud is all too easy in a busy department. When people are pushed to their limits in terms of high workload, they may realise that they can get something back from the institution to counter the amount of work they are doing. Items can go missing. Funds can be misappropriated and so can items purchased for and on behalf of your department. I do not want to dwell too long on this, but control is the issue here. If your systems are robust and you are monitoring them and challenging any discrepancies, you stand a good chance of avoiding this as an issue.

If fraud or theft is found within your department, deal with it quickly and decisively. Do not hesitate as you and the integrity of the department will be dealt a massive blow. Keep clean, and stay clean.

Making mistakes

Making mistakes is part of being human. You must try not to make too many mistakes, but sometimes they are inevitable. There is no Facilities Manager who has not made countless mistakes. It is by making mistakes that you learn how to amend your performance and move forward in a positive way. If you do make mistakes, own up to them and never try to hide them. You need to grow as a Facilities Manager and you need your integrity to grow with you. Start off with the right philosophy and you cannot go wrong.

Chapter 5
Outsourcing of services or in-house staff?

Reasons to outsource services

Outsourcing services can be a very emotive subject as you are dealing with the livelihoods of your staff. It must always be approached in a sensitive way and must be carried out under the scrutiny of your entire department. The choice to outsource a service or to use in-house staff must be made on the basis of sound rationale and should take into account the following issues.

- Does the service work; if it isn't broken, don't fix it!
- Does the service fit the requirements of the site?
- Is the service financially sound, or does it need a new financial model?
- Is there a sound reason for change?

Does the service work?

Do you like change? Be sure of one thing, your staff won't. Unless you have a sound, rational reason to tinker with a system, then leave it well alone.

Warning
Never approach change like a bull in a china shop; be methodical in your approach and plan well in advance.

However, a good manager always looks to improve the service that he or she provides. You should continually appraise your service to see if it still meets your operational requirements. We have already looked at how operational requirements can change through the application of good customer service. Change for changes sake is a poor way to develop your working strategy for your team, although trying something new can be good in small steps; if you don't test something, you will never know if it works. You need to develop a strategy to deal with these minor tweaks rather than approaching them in a cavalier way.

Outsourcing versus using in-house staff is exactly the same in terms of how you should approach it. You should not source any service in-house without first having carried out some background work to determine the type of service you are likely to end up with and whether this service would be better for your department and your client. Be careful. Once you start down the road of outsourcing your service, you may find that, if this is successful, internal politics may create a desire to outsource all your operations. Change is not always bad, but it needs to be embraced by all your staff from the start.

Does the service fit the requirements of the site?

Your goal must always be to provide the best possible service for the site. Only when you understand your site can you start to make decisions about whether the changes you are planning best suit the site. The hope is that through good, sound management practices your organisation will continue to grow. This is a natural desire of all businesses. With growth comes change, and this is likely to have a number of effects on your site and the service you provide. Things such as waste and recycling become much bigger issues that require managerial and strategic decisions to cope with and to keep within legislative requirements. You may have to start to add services which have never been needed before to ensure compliance, e.g. you may need a specific person to deal with waste or recycling. Or you may see a synergy between roles, combining roles to create a more streamlined service. Your site should come first, not your desire to create an empire. Any assessment of a service provided must be based on facts and figures, not on aspirations.

Is the service financially sound?

No Facilities Manager is lucky enough have an open chequebook for the provision of services. Your service delivery must be based on sound financial judgements and projections of likely outcomes. Although quality and service are important, if they are not financially sound in their application they will not be successful. However, sometimes services need to be provided that do not have a sound financial basis. These services are those that contribute to an organisation's image. These services need to be clearly understood as areas that benefit an organisation, such as window cleaning, plant displays or litter-picking. The image of an organisation should never be undervalued,

> **Warning**
> Service drift can undermine your service without you really knowing what's going on.

particularly in educational institutions. The "learner experience" is far more valued these days and may rate almost as high as the education itself.

This "learner experience" is an area where Facilities Managers can enhance their contribution by developing service. My background is in educational FM, so this book is sometimes biased towards education, but the same principles apply to all FM practices.

Is there a sound reason for change?

Sometimes issues come along that demand a change to the way you are doing things, such as financial austerity, which is currently high on the list of any government-funded agency or body. These boom or bust scenarios have been with us before and will no doubt be with us in the future. Sometimes it can be an organisation's response to these critical events that makes it stand out from the rest. Funding cuts sometimes enforce change and for the Facilities Manager this creates a challenge as to how to cut services while maintaining compliance. If a Facilities Manager truly understands his or her organisation and the services provided, then these judgements and decisions will be easier to take and will, in the long term, create a more robust organisation.

> **Note to self**
>
> When a business expands naturally, the services required will change. Always bear in mind whether you have the capacity to take on these new services, or whether you need to ask for additional resources.

If you are asked to reduce your budget and are expected to do this without it having any effect on the services you provide, then the first thing you are likely to feel is indignation. The assumption is that, if you can reduce your budget without changing the service, then you have been over-budgeting all along. The truth is, every now and again you need to re-evaluate your budgets and the services you provide to ensure you are not suffering from "service drift."

Service drift

Service drift is where the service you provide develops unchecked and becomes something completely different from the original service. A simple service can organically grow through success and become something completely different. Constituent parts of the service may develop, sometimes in an ad hoc way, to move the service on. This organic growth can sometimes lead to an over-complication of the service and duplication of administration. As a result of the way in which services grow, they may develop traits or good practice from other services. This is particularly true of administration. For example, you could have two independent services following similar "best practice" methods of administration, undertaken by two different people. It might be more practical and more economical to roll these two positions together into one post. This generates staff savings and may reduce the overall administration requirement of the organisation.

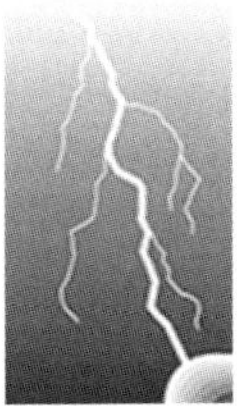

★★★★★★★★★★

List all the administration processes that take place in your department, lay them out side by side and look for similarities between them and any items that are duplicated. This can be your starting point for review.

★★★★★★★★★★

Care must be taken to make these sorts of decisions based on workload, not just on similarity of position. It is not practical to think that one person can suddenly take on the roles of two people unless there is capacity there and you can make genuine changes to your operation to reduce the overall administration workload in the first place.

We often over-complicate things, particularly processes, for little extra added value. It is these over-complications that we can target to make subtle changes to the way we work. Streamlining processes can sometimes lift a burden from staff, who may feel enlivened and released from unnecessary duties. We need to consider:

- Why do we do a process?
- What value does that process create?
- Are there any alternatives?

Note to self

Do we know if the things we do are worthwhile? Sometimes we do something because we think it is the right thing to do, but without it creating any real value.

When we began doing something, we had a valid reason for doing it. Or did we? Sometimes we do things because we feel we should. This is particularly prevalent in the public sector, where information and audit processes are so important.

However, unifying processes may take that requirement away. Sometimes we collate information and statistics on one system, only to enter the same data on another system. This duplication of processes may go unnoticed and you may find it has been a constant cause of irritation to the staff who have been implementing it. This kind of duplication is rife in most departments and can become embedded in their culture.

> **Note to self**
>
> Make sure you fully understand every aspect of the service being provided before you construct the specification. Essential items are: who provides the service; what is the service they provide; when do they provide the service; who do they provide that service to; and where do they provide the service?
>
> If you use this concept for all your contracts, you can focus on each one in turn.

Are you able to quantify the value of the process? In other words, do you ever use the information the process provides and is it really of any use? Again, you will probably find this has been a constant source of irritation to all your staff, but it has never been challenged.

You need to consider whether anybody has ever looked for alternatives to these processes or thought why you do them. Also, is there a better way of doing this? Sometimes departments never challenge why processes were started in the first place. If this culture is endemic, then you have a good chance of being able to make substantial changes without affecting the service provided.

Outsourcing: advantages and disadvantages

There are two types of outsourcing, each of which should be dealt with separately and which require a different approach. The first is simply buying in a contracted service from an external provider; the second is more protracted and involves market-testing an in-house function.

Outsourcing in its simplest terms is buying a service from the open market. To undertake this, you need to have:

- A formal specification of what is required, including service levels.
- A contract period, i.e. when the service will start and when it will end.
- The financial parameters of the contract, such as payment terms, how additions and deletions will be dealt with, the financial penalties to be imposed if the contract is not met and the bonuses awarded for successful completion.

Where the outsourcing is of a current in-house function, then this is far more protracted. You need to consider whether you are sure that this is what you and the organisation want. The decision to outsource should never be taken lightly. You should consider the likely implications of the decision and thoroughly evaluate the effect that this is likely to have on your organisation.

Warning
Be sure that the organisation has an appetite for outsourcing before you go down this route.

Specification is an art in itself; however, your specification is the most important consideration and, for the outsourcing of an internal service, needs to be watertight. It is important that the service you are specifying mirrors the service you are currently providing. For example, if you put in your tender document that post is to be delivered to a department by 9.00 a.m., you don't want to then find out that your staff have been delivering the post to sub-offices within the centre for the last year. This would have the effect of expectations not matching the specification, leading to amendment of the tendered contract and an increase in costs. You only need a few of these examples before you end up paying more for a service than you did before. You need to get your centres to buy-in to the concept of the new specification and encourage them to contribute to the specification process.

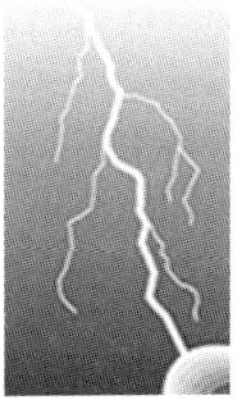

Remember that your staff are very important to you and if they are not fully on-board in any decisions on outsourcing, then negativity will build up within the workforce that may take years to put right. Your staff will be concerned about any changes and need to be part of the team that looks at the practicalities of outsourcing the service. Your trade unions also need to be part of the team to secure the rights of staff throughout the process. Staff consultation needs to be built in before putting a process out to tender.

★★★★★★★★★★

It is a good idea to try and list the benefits and disadvantages of each individual contract. This will help you focus on why you are outsourcing in the first place and help you see what you really want out of the contract in the long term.

★★★★★★★★★★

You may need to build counselling into the process. Never underestimate the pressure that the process of tendering an in-house service puts on the staff involved. This must be adhered to throughout the process. Although you must look at all times to the requirements of the organisation, never lose site of the fact that you are dealing with your staff's livelihoods and that a sympathetic approach will be recognised and appreciated by all. However, you must also be aware that no matter how much effort you put into staff welfare, you will never please all of the people all of the time.

Advantages of outsourcing

Outsourcing allows you to concentrate on other services while the outsourced service is managed by others. By freeing up your time from the day-to-day monitoring of the service, you can focus on the provision of other services. Outsourcing moves some of the risk from you, the client, to the contractor or supplier, who will have most of the responsibility for the service. It allows you to bring in specialist staff, which can be cheaper for certain areas of your service. Many larger companies that provide outsourced services in FM carry specialist staff who cover a number of contracts. They can then justify their specialism by economy of scale.

Warning
Always make sure you factor VAT into your calculations, especially if you are not set up to be able to reclaim it.

Outsourcing reduces your HR input, which otherwise can be a burden, particularly for services such as cleaning that may have a considerable number of operatives on site. Providing HR support to all these people is costly, so moving that responsibility away can save money. Outsourcing can guarantee the cost of a service if the contract is correctly drawn up; you can fix the cost of service provision for a number of years, which can help in uncertain economic climates. Outsourcing can provide a cheaper service, which is worthwhile, unless you are striving for quality as you main aim. In today's environment, the main driver is cost and outsourcing can help with this; for example, many catering companies will either update or put in brand new catering facilities as part of the initial contract.

Disadvantages of outsourcing

Outsourcing may end up costing more than the previous service if your specification is not tight enough, or does not deal sufficiently with variations. It is imperative to factor in your VAT costs, particularly if you are not able to claim these back.

You may also lose flexibility; your service provider will provide you with flexibility whenever you require it, but it will cost money unless correctly specified. Be careful when agreeing long contracts. Keeping your costs steady with a three-year contract sounds great, but if the service is poor and you have not written in robust termination clauses, you are stuck with a poor service for the duration of the contract. This can be draining on resources and draining on you.

Every change to the contract will require a variation. Never underestimate the power of the variation; even small variations can be expensive. Most modern companies providing FM services are masters at the interpretation of specifications. They will know at the outset what areas they can make profits on from variations and these will be exploited by the company to the maximum effect, even if you have been shrewd enough to include unit rates for these areas. However, this is what they are in business for, to make money, so you can't really complain.

> **Note to self**
>
> Make sure your specification contains the following key deliverables: what is the work to be done; where will the work be done; when will the work be done; and to what level of service will the work be done?

Your specification needs to be watertight, otherwise you have lost your control and will invariably cost your organisation a lot of money. You will pay a premium to cover risk; the risk factor of any contract is built into the final cost submitted by the contractor. The greater the likely risk, the greater the cost to your organisation.

Specifications

A specification is an explicit series of written requirements used to describe the level of service required in a detailed, yet clear and unambiguous, fashion.

A specification must contain:

- *Details of the work to be done.* Every element of the service you want provided must be included in enough detail to accurately convey your requirements.
- *Time frames.* The times at which the service is to be provided must be included, otherwise you could end up with service provision outside normal hours. You would then need to resource opening, heating and lighting a building unnecessarily.

- *Levels of service required.* In just the same way as you would write an SLA, your specification needs to detail the level of service required and the pinch points or target levels to be achieved.
- *How the contract will be monitored.* This should include how you are going to monitor progress and how you will evaluate compliance to specification on levels of service.

When writing specifications, some people fall into the trap of adding too much information on insurance, welfare, health and safety, payment and termination clauses, replicating this each time that they write a contract. In my opinion, the best way to deal with these areas is in a Prelims document. Prelims, or Preliminary Particulars, are the logical place to put peripheral information such as health and safety and payment schedules. These generally only vary slightly from one contract to the next, so why not create a standard document that can be tweaked for each contract depending on your requirements? This way, you know you have dealt with all the pertinent items without missing anything fundamental and you have saved yourself time by not rewriting standard clauses. A standard Prelim is given in the Appendices as an example.

Types of specification

There are basically two types of specification: prescriptive and performance. Prescriptive specifications describe the level of performance in very fine detail. For example:

> 4.1 Grass Cut 1
> Cut grass by cylinder mower to 25 mm at weekly intervals between mid-March and the end of October
> To include all follow-up mowing
> Note: the follow-up mowing to include all banks obstacles and any rough grass areas beneath trees within the area designated for weekly cutting
> Total number of cuts 33
>
> 4.2 Grass Cut 2
> Cut grass every two weeks by cylinder or rotary boxed mower to 25 mm
> Total number of cuts 16

This defines a height of grass cut and a frequency. This then becomes easy to monitor, sign off and control or measure performance. Payments are easy to quantify if unit rates are set up against each of the functions.

Prescriptive specifications can work very well and are not open to interpretation, but need to be accurate in terms of quantifying the workload. The work is described accurately and the overall process of monitoring is far easier. Prescriptive specifications have gone out of fashion to some degree in favour of the performance specification. For me, however, they still have a part to play, particularly where you can exactly quantify what it is you need.

The benefits of a prescriptive specification are as follows.

- You know exactly what you are getting; if you specify a particular piece of machinery, you know that you will get that piece of machinery.
- If you buy in quality goods, your long-term maintenance costs will be less.
- It is far easier to compare like for like services if they are specified in exactly the same way.

The disadvantages of a prescriptive specification are that it can cost more than a performance specification and it does not allow the contractor or supplier any flexibility in sourcing items and therefore waiting times for parts may increase.

Note to self

When choosing a performance specification you need to fully understand the service and to be unambiguous in your statements.

Performance specifications require a greater understanding of the task than prescriptive specifications and use a certain amount of subjectivity. They should be written with quality as the backbone for a positive outcome. The benefits of a performance specification over a traditional prescriptive specification are:

- it can transfer the risk for the design of the works or general administration of the contract to the contractor or supplier;
- it can allow the contractor or supplier to determine how to do things in the cheapest way and this may allow you to obtain the product or service cheaper;
- it allows innovation and invention into the process, perhaps leading to trailblazing results.

The disadvantages of a performance specification are:

- the contractor or supplier has almost free rein to provide whatever materials they need;

- when trying to source a cheaper way of doing something, contractors or suppliers could buy inferior or poor quality parts or services to complete the works more cheaply;
- if something is not defined fully, how do you monitor and manage the process when interpretation does not allow definitive assessment?

Examples of a cleaning specification and a grounds maintenance specification are given in the Appendices.

Risk

What is risk in relation to a contracted-in service, and how does it affect your organisation? Risk, in general, is rarely factored into contracts by the client and is sometimes over-factored in by the contractor. Risk is a risky business! In terms of contracts, risk can be described as "a potential adverse effect, either quantified or unquantified, which has the possibility to either create a detrimental condition or destabilising effect on the contract or those involved in the contract." So what kind of risks can a contract have? These can be classified as financial, legal and disruption of service.

The main financial risks are as follows.

- The prime objective of a tender exercise is to purchase a product or service at the cheapest possible price; however, it must be understood that this is not always possible and the risk is that you could end up paying more for a service than you did before.
- If you get things wrong in tenders and contracts, you get them wrong for the whole period of the contract. That could be years, unless you have written in break clauses.
- If you have specified things incorrectly, then your costs can spiral out of control.

With regard to legal risks, most contracts contain contractual and legal provisions which, if not adhered to, can put both the client and contractor at risk. In particular, the implications of the Transfer of Undertakings and (Protection of Employment) Regulations 2006 (SI 2006/246) can put an additional risk on the contract that would not be there if in-house staff were used.

The following risks are associated with a disruption of service.

- If sourced poorly, a contract for the provision of services can cause a disruption to services. If services fail due to incorrect specification, then this presents an undue and unnecessary risk to the organisation.

- Even if you have made the contractor financially responsible for breakdowns, the disruption to service from the breakdown of poor quality parts will live long in your staff's minds, even when the original problem has been solved.
- If a specification drives down quality by putting a higher priority on quantity, then the level of service provided will be lower than that required by the organisation.
- Under-resourcing a function can affect both the quality and the overall provision of the service and may mean that the service does not reach the desired levels of SLAs.

The education sector: the "learner experience" and outsourcing

What is the "learner experience"? As far as the FM department in an educational environment is concerned, it is one of two things: how the students interface with the site and how the students interface with your service. In other sectors, you could substitute the term "student" with "visitor" or "customer" and the "learner experience" with the "visitor/customer experience." Both will give valid and vital areas for you to look at.

The site, in terms of the day-to-day activities of students, is secondary to the level and depth of the teaching experience; however, many of the judgements students will make about the organisation are based on how it affects them, such as:

- are the buildings warm and inviting
- are there suitable places for leisure, e.g. common rooms and cafes
- does the layout and signage of the buildings help the user to move around them
- are the classrooms conducive to learning
- does the site feel safe and welcoming
- are the toilets clean?

All of these influence the students" outcomes, sometimes in a small way, but sometimes significantly. Students have been known to leave college because it is cold or they do not feel safe. If a negative image is purveyed by the college, this will percolate through and affect the learner outcome. The services you provide can also influence the learner. If your staff are negative, then again this is likely to have a negative effect on learner outcomes. This is as much an issue for you as a Facilities Manager as it is for a Curriculum Manager. No one wants to be responsible for a student's bad experience.

Sometimes buying in a service can alienate it from the site, as it no longer feels a holistic part of the provision. This can result in bought-in services seeming negative and uncaring towards the students. This, in turn, can have a negative effect on the way in which the service is provided.

The learner journey is often seen as a holistic process. From the moment a student sets foot in a college, they are forming opinions about the organisation. If they do not feel confident about the professionalism of the college staff, you can understand why they would not feel confident that the college is providing a good overall outcome. This could ultimately result in poor grades for the student or, at the very least, will give you a negative score on your QDP returns, which could undermine the level of service provided. It is therefore important when considering your site as a whole to also consider how your services will interact with the students. This is an example of customer service, so whether the customer is a student, visitor or consumer, the same principles apply.

> **Note to self**
>
> It is vitally important to check the quality and competency of the contractor's service before engaging them. Afterwards is always too late and you will be criticised for making poor decisions if a service fails.

Working with contractors

It is important to understand that you cannot economically provide in-house services to cover all the requirements of a busy site; at some point you will need to bring in contractors to fulfil certain functions around the site. The amount of additional specialist contract needed will depend on each individual site. Contractors should be used for specialist areas bound by specific skills or legislative requirements, such as heating engineers, or in areas where it is uneconomic for you to provide that particular service.

If you are going to use contractors, then you need to know the implications of this with respect to your duties as the procurer of that service and their duties as guests on your site. You need to know this before you start looking at what the contractors are doing on site and the specification of these duties. You are responsible for the contractors while they are on site and you can be held accountable for their actions.

Choosing your contractors

Choose your service providers carefully and build competency into your tender regime. Your contractor should be competent to provide the service. Make sure they are registered with an appropriate professional body for the function they are providing for you, e.g. gas engineers need to be Gas-safe registered and ventilation workers need HVCA certification. It is not just the professional bodies you should be looking to when using contractors; registered bodies such as the Contractors Health and Safety Assessment Scheme (CHAS) and Construction Line will provide a certain amount of surety that your contractors are competent to undertake the works. Speak to other local organisations and find out who has worked with them and what they thought of them.

Make sure your contractors are carry the relevant and necessary insurance and that they are insured to a sufficient value based on the size and complexity of the operations. Ensure that you obtain copies of both their personal and employer liability insurances and make sure they are current.

In an educational environment where the contractors may interface with your students, some of whom may be vulnerable, make sure that the individual operatives who come on site are covered by a current Disclosure and Barring Service (DBS) check. Try to encourage your contractors to keep a core of staff who come to your site. This makes sure you comply with the CRB requirements and also allows a relationship to build up between the contractor and your department. This relationship makes your life easier as problems are dealt with quicker when an operative or engineer knows the site and can identify problems. This is equally important in a retail or health environment.

Make sure the company is trading in a favourable financial position; you do not want a company going bankrupt midway through an important contract. If a company is fully solvent they are able to procure materials internally, thus providing a better service to your team.

The company must carry the necessary health and safety skills for the works needed and have a proven track record in these. Ask them to demonstrate this by providing a copy of their health and safety policies and procedures. You will soon gauge from the quality of this information and how it is completed if you are likely to have problems in this area. Health and safety compliance is of paramount importance when works undertaken by contractors could affect students and staff. Your contractors must not cause unsafe conditions on site which put your students at risk. It is preferable to have the same team working on site at all times; from a health and safety perspective this provides you the knowledge that contractors are conversant with your equipment and aware of your procedures, such as hot work permits.

Contractors are increasingly being asked to report on their sustainability performance. Clearly this is an area which should tie into your own strategy on sustainability. Sustainability, as far as contractors are concerned, means: how and where they procure their parts and materials in terms of distance from site; are those parts and materials from sustainable sources; and how does the contractor deal with their waste? Sustainability is also about whether they set targets for sustainability each year. Are they achieving those targets and showing an annual improvement on their sustainability performance?

All this information, once collected, should be reviewed on an annual basis. Set up your administration procedures to reconcile your contractor information on an annual basis, requesting details of insurance and updates on sustainability and health and safety performance. This is easily done by setting up reminders in your Outlook calendar. Don't forget you need to review to check compliance and completion and to link this in with the quality of the finished service. Do not forget that the drivers for quality used to measure your own staff should also be applied to external contractors. It is no good getting a cheap service, only to find the quality is poor. Build quality into every process and you will ensure you move forward in a proactive way.

Performance and review

Performance and review of your contractors are essential. Why pay good money for a service and then not check if you are getting what you paid for and obtaining good value for money? Work out a common methodology for scoring your contractors on the essential areas of competency, insurance, safeguarding, solvency, health and safety, sustainability and performance. Immediately you will have a set of criteria by which to measure your contractors against each other. Setting this up on a spreadsheet, setting up a scoring matrix and then setting a minimum acceptable score immediately gives you a methodology for assessing if you have sufficient quality contractors on your selection list and whether you need more contractors to give a high level of quality and allow your senior management team to see that you have shown due diligence in procuring their services.

Warning
Avoid buying services simply because they are cheap. You are certain to get a poor service as a result.

Quality

Do not accept contractors simply because they are cheap. You need to be constantly looking to engender quality into your processes and this means that your contractor lists should be subject to the same scrutiny. Quality is not always obvious in relation to contractors, particularly when taking on a new contractor or supplier. In the early stages of procuring new contractors you will find that they try to appear to be the most professional company in the world and will promise you anything to get you to commit to a contract. This does show a capability to perform, sometimes in a quality fashion, but you need them to provide consistent quality throughout the whole life of the contract, not just at the beginning.

The best way to ensure a consistently high level of service is to obtain references from previous clients. Don't fall into the trap of asking for references from the company itself, because you will only get contacts from clients with whom they have had superb working relationships. Try to find out where the contractor provides a similar service locally and then make an unsolicited call to the previous client. Try to contact a similar institution to your own. Remember, as Facilities Managers we are all in the same boat and colleagues will usually be only too happy to report poor performance from contractors.

Any sizable contract such as cleaning or security should show collective team responsibility (CTR) in the provision of the service. This means they operate as one team, not as fragmented sections. They should act with a single purpose and not provide a variable service level on site, showing an overall team spirit and a desire to do well. Most of all, they should accept responsibility for the service and not expect you to continuously monitor performance and quality.

What should you ask for when speaking to potential referees? Generally your questions should cover the following areas.

- Do you get value from the service provided? Are variations accurate and not inflated and are invoices on time and accurate? Does the contractor seek variations proactively or are they prepared to negotiate and be flexible to suit the service?
- Is the service provided on site friendly and does the team buy-in to CTR? Does it reach the specified level of quality? Does it have a methodology of constant improvement and the desire to self-improve? Does the service engage with students? How many complaints have you had about service provision? How many positive comments have you had about the service?

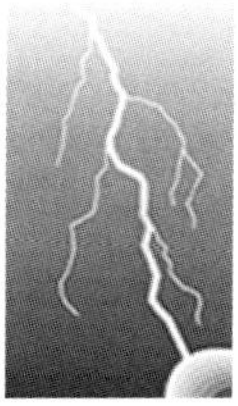

★★★★★★★★★★

Periodically undertake a spot check on the items going into your skips and front-end loaders on site. You may be surprised to find they are full of waste from your contractors.

★★★★★★★★★★

- From a management perspective, does the service appear to be run well? Does the management team attend regular feedback meetings? Does the on-site team rate the off-site head office function and, if not, why not. Does the team follow a QMS?
- From the works perspective, is the finished job completed well and on time? What are the levels of snags requiring rectification? How does the company respond to snags? Have they used quality materials? Do you have a lot of breakdowns (for M & E service provision)? Are your staff happy with the finished job? Sometimes you can glean a lot from asking staff what they feel about the quality of work. Are the contractors on site when they say they are?

I'm sure there are many more questions you could ask. The key is never being afraid to ask those questions before you enter into the contract. Once you have signed a contract you put yourself at a disadvantage and will end up doing far more work in the long term.

Waste

> **Note to self**
>
> Pull together a standard set of questions to ask referees and use this to ask the same questions about each contractor. This gives uniformity and allows a consistent scoring matrix to be used.

The contractor will provide waste, just the same as if you were providing the service yourself. This may be in the form of boxes or may be grass cuttings from the gardening team. You must make it clear in your contract that the contractor must remove all waste from the site. If you don't do this, you will end up paying for the disposal of the waste generated by the contractor as they will just put it in your skips and front-end loaders. Make sure that this process is auditable and that you are able to see how the waste has left the site and how much has been removed. The best way to do this is by checking the waste transfer notes from the contractor.

> **Note to self**
>
> You need to find out who has used this contractor before you and how successful they were. Speak to your colleagues in other organisations.

There is, however, some waste from the contractor that you need to see to prove that they have completed the work, e.g. parts removed from machinery or kit on site; otherwise, how do you know they have replaced the defective parts? The best way to do this is to have a waste parts box in each of your plant rooms. You can then see what has been removed from machinery and what has been repaired. You can ask them to empty this box and remove it from site at regular intervals.

Health and safety

Health and safety is definitely an area where you need to spend time making sure that your contractor or supplier complies with provisions while they are on your site. It is your duty is to supervise and check your contractors while on site and you have an obligation to look after the welfare of everyone who comes onto site; this means contractors as well as your own staff. Start as you mean to go on.

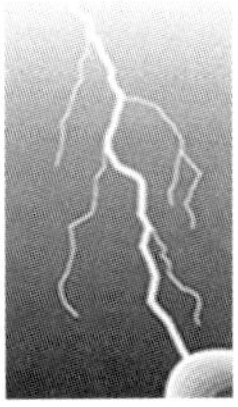

★★★★★★★★★★

Set up a standard health and safety checklist for each contract and append this into the contractor's file.

★★★★★★★★★★

Checking your contractor for health and safety performance is the key to starting off on the right foot. This must be done before the contract starts to allow sufficient time to undertake the checks properly. A health and safety Pre-Qualification Questionnaire (PQQ) is included in the Appendices to help you. These PQQs are not just pieces of paper; they need to be looked at in detail. If you knowingly take on a contractor who has a poor health and safety record and you have a serious issue on site, do not be surprised if you have to take responsibility for their poor performance. By unquestioningly accepting a PQQ from a contractor you make yourself culpable. Make sure you:

- Send out a health and safety PQQ that covers your liabilities and identifies poor performance clearly.
- Check the documents when they are returned; don't simply file them.
- Act on any inconsistencies in the contractor's return and challenge any seemingly poor returns.
- Do not take on a contractor who performs poorly on health and safety unless you are sure they can comply fully with your requirements and undertake the service you are requesting safely.

Compliance with health and safety issues can be part of contractual compliance. Many of us use Prelims for every contract. Generally these will detail the health and safety requirements and should point the contractor in the right direction in terms of compliance on site. Contained within any good set of Prelims will be how the contractor is to monitor and manage health and safety while on site. If this is in your Prelims, you also have a duty to ensure that the contractor is adhering to those requirements. The best way to do this is for you to have health and safely as a formal agenda item at every site meeting. This way you can pick up early non-compliance and deal with it promptly. Also make sure that you include health and safety in your SLAs, with set targets to be achieved. This way the contractor will have to prove compliance to receive payment, taking some of the pressure off you.

Contractor files

It is important that you are able to control your information about every contractor who comes on site. The most useful way to do this is to set up a contractor file for each individual contractor. This will allow you to record details and store them in an easily accessible way. This can be a paper-based file or, better still, a file held on computer. This will enable you to quickly check and monitor whether you have up-to-date information on public and employer liability insurance, health and safety performance and CRB checks. It also proves from an audit perspective that you are compliant in your approach to the adherence of contractors to health and safety legislation.

This file should be distinct from the contract file as your contracts and contractors could change. You will need to keep your records on health and safety for a number of years, so keeping this information separate from the contract file will allow you to archive records not currently in use. There are examples in the Appendices of documents dealing with health and safety for contractors, including a health and safety questionnaire.

Procurement routes for FM

There are a number of tried and tested procurement routes for FM. The choice of route will depend on the service you are outsourcing and the complexity of the service. Much will depend on you knowing exactly what the service is and having the ability to put these definitions accurately into a specification. You need to know what you are trying to achieve, whether this is financial reductions, quality improvements, or simply the ability to undertake specialist functions outside the skill set of the organisation. The procurement route may also be influenced by the size of your organisation, its complexity and its geographic spread. The procurement routes available in the FM sector are described in the following Theory and information slots.

Theory and information slot

Single service procurement is the purchase of individual services as and when required. There is no interlink between the service you are buying and others within your portfolio.

Benefits of this type of procurement include the following.

- It is a simple process which allows you to make purchases in bite-size chunks.
- It does not require integration between other services.
- Services can be procured over a period of time; purchasing a few services at the same time allows you to smooth your resources over time and to phase procurement.
- It allows you to choose the best individual services rather than a bundled pack of services that may have varying levels of performance.

Theory and information slot

Bundled services are when two or more services or functions are provided by the same provider.

Some of the benefits of adopting this approach include the following.

- A fixed price can be agreed for services.
- Risk can be moved from the client to the contractor, particularly in specialist areas such as heating and HVAC.
- Staff and multi-skilled workers can be optimised.
- Lower contract prices can be obtained because of economies of scale and sharing of administration.
- Specialist services that the client would find difficult to provide in a cost-effective way can be pooled.

Theory and information slot

Integrated FM is the bringing together of several support services into a single management structure. The adoption of integrated services allows a collaborative approach between the supplier and client, while allowing streamlining of the management structure.

The benefits of integrated services are: a streamlined structure; cost savings; innovation between integrated services; shared best practice; and the viewing of service provision in the round, encouraging collaboration.

Theory and information slot

Total Facilities Management (TFM) is a single source of supply for all your FM needs and is an ongoing partnership with the service provider. All services are provided or bought in from an external provider and monitored by a client manager.

Some of the benefits of adopting the TFM approach include the following.

- A fixed price can be agreed for services.
- Risk can be moved from the client to the contractor.
- Staff and multi-skilled workers can be optimised.
- Lower contract prices can be achieved by economies of scale and sharing of administration.
- Management is streamlined by having only one point of contact.
- If applied correctly, a clear definition of service can be achieved.
- There is the potential for reduced administration cons on the client side.
- On larger contracts, services spread across the portfolio can generate economies of scale.
- TFM contractors have a vested interest in innovation and continuous improvement.
- The condition of buildings and stock can be more accurately defined and monitored.

Theory and information slot

Total infrastructure FM (TIFM) is mainly used in larger organisations with both wide geographical spreads and diverse portfolios of building stock. It allows widespread stock to be compared and contrasted, while sharing best practice and creating a database of information and strategic tools for the organisation.

TIFM gives the following advantages.

- A reduction in overall cost by accurate tracking of the performance of stock.
- A more reasoned strategic view of operational needs based on facts.
- Integrated systems allow workload to be spread and better centring of costs.
- A reduction in capital expenditure; with the ability to accurately track resources comes the ability to see spare capacity, allowing reallocation rather than capital new build.
- Being able to visualise the portfolio allows you to see both anomalies and best practice.
- Accurate data allows you to visualise your portfolio and to use statistical information to show performance in a graphical format.
- TIFM can legitimise FM as it moves from being a functional to a strategic service, allowing the true worth of FM to be shown as a benefit to the business and not simply a support service.

Aide Memoire: Outsourcing of services or in-house staff?

- Have you evaluated whether the service provided fits the site's requirements and meets the strategic objectives of the organisation?
- Have you looked at and documented the advantages and disadvantages of outsourcing versus keeping an in-house workforce?
- Have you looked at the specification to ensure that it is fit for purpose and meets the requirements of the site?
- Do you know how to evaluate your contractors?
- Have you chosen your procurement route?

Chapter 6
In-house teams

Maintenance team

The term Facilities Manager has been around for a long time, as has the term Maintenance Manager. The term maintenance has stayed with our role and, irrespective of whether you are a Facilities Manager or an Estates Manager, or a combination of both, people will always associate you with maintenance. It will generally be the first item on your job description and other members of staff will look to you as the person who arranges to change their light bulb when it goes out. Maintenance is rarely understood, even by many Maintenance Managers. We are much more than light bulb changers. We have complex roles and juggle multiple disciplines to fit the 37 (or probably closer to 60) hours of our working week. However, to be successful in our chosen trade, we must be maintenance specialists. This is one of the cornerstones of our job and the area we are expected to excel in.

Your maintenance team should never be more tuned into maintenance than you are. They will run you ragged if they think they know more than you. So you must be proactive about your maintenance team.

The size of your team will obviously vary between organisations depending on their size, location and available budgets. Team size should be tailored to your requirements and aspirations and dovetail into all the other teams on site. Maintenance teams generally have a hard time because of preconceived ideas about builders. Modern builders are no longer like those of the1970s and 1980s, but some stereotypes seek to unravel any hint of professionalism before it has a chance to flourish. Sometimes

Warning
Try to break down the stereotypical image of the builder. Tackle the team before tackling the image held by staff.

this stereotypical image is self-perpetuated by the team itself in an aim to look cool, but there is no real place for that sort of image in a modern organisation. This image does not fit the modern ethos of safeguarding, or the moral responsibility for equality and diversity. This is the first issue you will need to tackle in your maintenance team. However, the personality of the maintenance team rarely differs between organisations and they are always a friendly and jovial, if sometimes cheeky, bunch and this friendliness endears them to staff. Do not remove this trait, or your customer service levels may fall.

Specialisms

Unfortunately, the days of having specialists in each department have long gone and the one-man, one-job philosophy is also very outdated. You still need those dedicated specialists for roles such as electrician and heating engineer, but, if not economical, these services can be bought in. You are usually looking at a team consisting of a plumber, joiner, painter and perhaps a groundsman. You should try to make these trades interchangeable to obtain maximum flexibility and to allow your staff to learn the skills of the other professions in the team. Developing a multitasking team will create a good bond between team members and provide sickness and holiday cover without too many problems.

> **Note to self**
>
> Get to know your maintenance team, understand what skills are available and try to encourage a multi-skill approach by staff.

Workload

It is dangerous to just assume that there is there is enough work for the team. You need to be able to confidently detail what work your team is undertaking and be able to demonstrate that they are functioning at a notional level of performance. This can only be done by tracking the work that has been carried out. Generally this is achieved via the Helpdesk, where the jobs are issued. If you don't have a Helpdesk, then the task becomes far more difficult.

At the very least you need to be able to track the following areas.

- how many jobs are being undertaken by each trade
- where those jobs are being undertaken
- performance, e.g. how long it took to do the job

- materials used
- special arrangements made
- potential costs

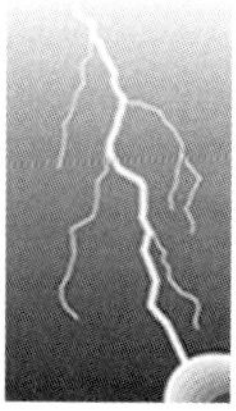

★★★★★★★★★★

Start to break down what your maintenance team does by quantifying all the elements, such as time, volume and materials. This will provide you with an auditable system.

★★★★★★★★★★

It is important to recognise that the moment you start checking how much work is being done by your department, you will immediately open the floodgates of childish behaviour to try and prevent you checking up on them. It is human nature for staff to not want to be monitored; it does not necessarily mean they have something to hide (see the earlier discussions on Theory X and Theory Y workers in Chapter 3).

Why do you need to check the workload? There are a number of reasons why you need to know how many jobs are being undertaken.

- It allows you carry out like-for-like comparisons for each trade to give an overview of the productivity of individuals.
- It allows you to plan future works by comparing trends from previous years, e.g. how many jobs are carried out over the summer break, allowing you to plan round larger projects while maintaining cover for regular works.
- It allows a month-on-month view of levels of work, allowing you to see dips in volume; this can help you plan holiday breaks, for example.

Where are these jobs being undertaken? Again, there are a number of reasons for needing to check this.

- If you want to counter-charge departments, you need to know which jobs are being carried out in each department.
- It allows you to carry out comparative analysis on each building to see which are the biggest drains on your resources.
- It allows you to see where the failure rate is increasing in individual areas, which could influence your planned programme, e.g. an increased level of boiler failures may indicate the need to replace equipment.

Performance

It is not simply enough to see how many jobs are being undertaken by individual members of your team.

- Some jobs may be quicker than others, e.g. joinery jobs could be quick five-minute jobs such as easing a door or fitting a new lock barrel, whereas painting a whole room will take longer; recording the number of hours worked is very important.
- The total number of hours worked allows you to see where future resources need to be targeted.
- Monitoring performance can give you ammunition when the workload seems to outstrip the available resources, allowing you to justify additional staff.
- Monitoring can also allow you to see where staff are underutilised, not so that you can get rid of them, but to enable you to broaden their individual roles to strengthen the department.
- Records also need to be kept of materials used.

Special arrangements cover areas such as health and safety (working at height or hot work) and allow these to become part of your auditable tasks. Linking these tasks to jobs is your double-check that health and safety procedures are being adhered to.

It is important to be able to cost-centre works. It makes good practical and auditable sense to know exactly where your materials are going. In this way you can be sure that they are not disappearing off-site; unfortunately, this does happen. In terms of stock control, you need know exactly when you need to replenish your stores. You can also track where the materials have been used.

> **Note to self**
>
> Never forget to factor health and safety requirements into your working calculations. This is one of the most important areas to be able to account for in an environment where visitors are present and accidents can easily happen.

Knowing how much works cost is becoming increasingly important. Global costs will give you the running total for the site, but knowing what each individual building or discipline area costs allows you to make strategic judgements on buildings and trades. You can then establish and account for the service into the future. Knowing your liabilities is the key to understanding your site.

Contracting in or out?

We have looked at insourcing in the previous chapter. Your maintenance team is one area where you may have a dilemma as to how to best provide the service. There is no doubt that you need service provision; things will always break and need repairing and equipment will always need to be serviced and maintained to prevent breakdown. However, how best to undertake this service is a debatable area.

The only sure way to work out the viability of service provision from an in-house or external perspective is to have an unambiguous and detailed assessment of the level of work on site, tied in with an appreciation of the cost of buying those services in. Your assessment of levels of performance can help you to make this judgement on outsourcing viability.

Your assessment on viability cannot be a simple figure pulled out of the air. It needs to be based on factual evidence. Your Helpdesk should provide you with details of the volume of work, the cost of that work, timings of that work, and success rates, e.g. whether it was completed on time. This is the basis for your assessment. However, you need to layer over this a number of other factors.

- *Customer satisfaction.* How well does the service perform its duties and does it meet all the criteria from the client in terms of completion without disruption and to the satisfaction of the work area?
- *Response.* This is the main reason most organisations plump for in-house services, because they undertake jobs quicker and more effectively. Response is usually the key selling point for an in-house service.
- *Cost.* The single biggest driver for most services. There is an assumption that in-house services are cheaper. Are you able to quantify the true cost of your service provision and have you included staff costs (including pensions and National Insurance, training), materials costs (sometimes these are lost in other budgets and not always charged or assessed as the total cost of the job), property costs (every team needs accommodation, storage space and delivery areas and insurance also needs to be considered), health and safety costs (sometime huge back-of-house functions exist to support an in-house team), HR and administration costs (again a back-of-house provision) and procurement costs.

You can see what a phenomenal task just costing your service can be and, the bigger the department, the bigger the task. Armed with your assessment of the volume of work, the service levels needed and the cost, you will begin to see the true cost of the service. You can then relate this back to cost per member of staff, or cost per building or per square metre. The options are as great as you choose them to be. Your assessment may then be sufficient, without doing further work, to identify:

- total jobs undertaken in each building or for the site as a whole
- total cost of providing those services
- details of service levels and prescribed service level indicators
- a workload appreciation throughout the year, taking into account the different academic terms and holiday breaks
- breakdowns by trade giving resource requirements
- how all of these tasks link to your strategic objectives

These will give you the necessary details to construct either your prescriptive or performance specification. This will allow you to build up an accurate estimation for going out to tender. Remember, the more accurate and detailed the specification, the more likely it is that you are going to get an accurate and realistic tender return.

Site supervision team

This is the most visible service you provide, yet it is the most elusive to quantify in terms of what your staff actually do. Site supervisory services are endless in terms of definition. In true FM spirit, these staff do everything that nobody else wants to do. Hence your site supervision or caretaking teams can be dealing with:

Note to self

Make sure your site supervision section has a full set of directives and tasks to undertake. If these are not identified fully, they will always be seen as dispensable and will be the first area targeted when there are cutbacks. Make sure they have a quantifiable and measurable role.

- locking and unlocking buildings
- set-ups for room changes
- moves and relocations
- janitorial duties, such as unblocking toilets and cleaning
- general porterage

- post and deliveries (if you don't have a post room)
- site security, over and above that done by your security team
- customer service, such as ferrying people around and guiding visitors
- health and safety, e.g. access and egress maintenance and helping people out of broken lifts
- minor repairs, such as putting up shelves

The difficulty with this wide range of duties and expectations is that this service can sometimes be seen as superfluous and be the brunt of any cuts needed within a department. You should aim to keep these staff as they can be worth their weight in gold.

This team can sometimes disappear into the background because of this inherent difficulty in defining what the service is. Your challenge is to quantify what they actually do, so you can justify the service when a challenge comes your way. Every function undertaken by site supervisors or caretakers is a job. These jobs should be logged and quantified in the same way as your maintenance jobs. How else are you going to quantify the workload, set benchmarks and measure the service? It is also the easiest service to leave alone to get on with its work in the background. This is a simple mistake to make. Don't assume that the work undertaken by this team is inconsequential; people would soon object if it was suddenly withdrawn as a service.

Your site supervision team, along with maintenance, is the team which controls your PPM process. Carefully planning PPM to be undertaken by your site supervision team can justify the team and give them something concrete to undertake, which influences your building maintenance strategy.

Your site supervision team are perfectly matched to PPM tasks such as:

- *Legionella* testing, e.g. hot and cold water temperature tests and running off infrequently used outlets
- window cleaning in key areas (but not at height)
- access and egress checks
- oiling of door closers and hinges
- external light checks
- room reports (checks on safety issues and wear and tear)
- sustainability checks such as turning off lights and PCs

These should be logged centrally on your database of PPMs and actioned, monitored, documented and audited.

Timing of service

Your site supervision team will be the team that spends most time on your site and will probably cover from 6 a.m. to 10 p.m., with occasional additional weekend shifts. It is therefore important to plan an even flow of times on site, or at least one which is matched to the workload. Matching your staff to your overall workload will prevent you accruing large overtime bills. A simple mapping exercise outlining your organisational requirements will show exactly when your service is needed. Don't fall into the trap of trying to cater for services that you yourself provide, e.g. opening at 6 a.m. to accommodate a cleaning regime. Why not start your cleaners later and capitalise on not needing to bring your staff in early to cover? You also benefit from not needing to heat and light the buildings early in the morning and this will help you with your sustainability objectives.

Warning
Make sure your team is fully trained as an accident could be expensive in terms of compensation claims.

Integration

It is sometimes a missed opportunity not to look at integrating services to get best value out of the service provision. This is not wholesale amalgamation of services, but using them wisely and using the strengths and weaknesses of the service to best effect.

A good example of this approach is looking at the synergies between services. For example, your site supervisors fulfil a security role because they can keep an eye on insecure areas and intruders on site and report these back to the security team. Your cleaning staff can also be your eyes and ears as far as security are concerned. Using all three teams holistically should give you a better coverage of your site. The enhanced security offered by taking account of all three tiers of security makes your risk assessment more robust.

Another example of integration is opening up and closing down procedures. If you

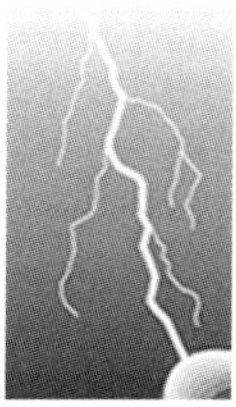

★★★★★★★★★★

Map out all the tasks undertaken by your site supervision teams and try to make connections or synergies between different jobs, e.g. which jobs could be carried out at the same time, such as parcel and reprographics deliveries.

★★★★★★★★★★

integrate your site supervisors and cleaning regimes to come together during locking down procedures, e.g. getting your cleaners to lock down areas as well as your site supervisory team, can help to stretch your resourcing of that function. You can them use some of your staff during the day on more productive items such as PPM. You must, however, be careful not to transfer staff duties from one team to another without first considering the potential for fair and equal pay comparisons. By changing one job slightly, you may end up making the job content similar to that of another role. If one group is paid more than the other, you could get a legitimate challenge under fair and equal pay that you would not be able to counter.

Think outside the box and try to pinpoint potential synergies between services so that when you need to reduce costs, you have a fallback position that does not compromise the safety of the site and the service you provide.

Safety and security

We have touched on this already, but there are other functions that your site supervisory team can provide and integrate into.

- *Fire procedures.* Use your site supervision team as your fire marshals throughout the site. This is one of the key functions of the facilities department; staff will look to your team to take the lead in fire strategy and management.
- *Bomb threat response.* Use your team to sweep buildings for suspect packages. Increasingly, your team need to adapt their fire safety procedures to suit the added burden of bomb threats and intruders. This is a key role that needs to be undertaken by your team. An example of bomb threat policies and procedures is given in the Appendices.
- *Rescues.* Your team can help people out of lifts which have jammed, or use Evac chairs during fire practices.
- *Alarms.* Your team can activate and silence alarms.

Resourcing

Your site supervision team is a bit of a mop-up service in terms of getting work done which does not fit into a specific niche. Of all the services you provide, your site supervision team needs to be the most flexible in both its approach and its capability. The staff you employ need to be the most flexible and should be encouraged to take on additional tasks. This flexibility may create the misunderstanding that they don't really do a lot.

It is easy to slot this team into functions that are difficult to otherwise resource such as PPM and sustainability. Sustainability is one of those areas where they can really make an impact, undertaking works such as the following.

- turning lights off
- checking heating has gone off in accordance with timings on the BMS
- checking windows are closed
- taking thermometer checks of areas to ensure compliance and to justify heating adjustments
- carrying out external checks on lighting in public access areas
- checking whether classes are on to allow you to make recommendations about early building closures
- recycling duties such as sorting waste

Core times and resourcing

Core times are sometimes difficult to establish, particularly in a college or university environment. The curriculum day is getting longer each year as colleges become more flexible in their provision, so it is not unusual to have a college open from 7 a.m. through to 10 p.m. It is obviously not practical to fully resource with equal numbers of staff for the full day, e.g. if you have eight site supervisors, logic tells you that you need two shifts, one a.m. and one p.m. to cover the full day, with four on the morning shift and four on the afternoon/evening shift. This may work for you, but you need to know when your core work is required. In an office or manufacturing environment, your core times may be far easier to establish and track.

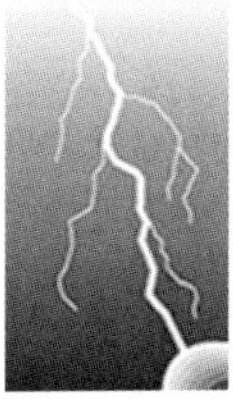

> Look at when you need to provide services on site and try to match your structure to the times required, or you may find your overtime bill goes through the roof.

This can be plotted if you understand your workload through schedules, work calendars, or the Helpdesk. You can easily become unstuck when resourcing your days because your employees may not advise you that you have too many staff working and you are not usually around in the evenings to monitor and manage their workload. Supervision is the key issue.

There may be a valid requirement to keep the site open for the maximum amount of time to allow the organisation to provide the flexible service provision required in today's market. This is an essential part of the site supervisory aspect of your service and needs to be dealt with and covered proactively. Your job is made harder by having to stretch resources to cover these changes. Of course, you could choose to cover the additional services by throwing overtime at the problem; and your staff will let you do this. Overtime should only be a temporary fix. If you need to commit large amounts of funding to overtime, your schedules and rotas are wrong and you need to think about restructuring. This situation is likely to get worse, not better. As a manager, you should be trying to think of practical ways to resource the changes without resorting to overtime.

The "friendly" service

Your site supervisors are the face of your department, sometimes even more so than the maintenance team. They are seen by most people as the first point of contact for FM. This is good, as you will find that your site supervisors build up a rapport with staff at every level and they can be helpful if things go wrong between your department and others within the organisation. Genuine feedback from your staff on the mood of other departments should be encouraged as it works for you, them and the other departments in a positive way. Customer service is very important for these frontline staff and training on customer service techniques will never be wasted. Presentable staff dressed in clean uniforms help to present a professional image. Making sure your team look right is vital in maintaining the identity of your department.

> **Note to self**
>
> Check out what happens on Friday evenings! Are you resourcing something when there is no one in the buildings?

Helpdesk team

A Helpdesk should be at the centre of every FM department. It should be the channel for all work, both in and out of the department. The Helpdesk should be: the hub for the entire department; the place to report any defects or problems on site; a channel for work, e.g. issuing of works; the customer service department; and the audit department.

When used correctly and linked into all aspects of your work, the Helpdesk should effectively allow your department to become a fully customer-facing operation. In some smaller businesses the Helpdesk duties may need to be rolled into the administrative function, as sometimes it is not always viable to allocate specific staff to the function of Helpdesk. Again, look for synergies between jobs. It could be that in its simplest format the Helpdesk is the maintenance team.

The hub for your department

Your Helpdesk is the hub of your department and should be the main interface with your customers. With this in mind, it should be located in a position where all staff can access it and see that this is the place to go to resolve all FM-related issues. It should also be the key area that FM staff themselves report to and liaise with to receive their workloads, report on progress, problems and completion and other work-related issues, and report follow-on or repeat works.

The place to report defects or problems

Location is the key to the success of the Helpdesk. If it is not visible, your service becomes a faceless department. Facilities Management problems can sometimes get out of control, or seem never to be resolved, because people can find it difficult to articulate problems accurately without that face-to-face friendly contact. A friendly service creates empathy between the client and the Helpdesk staff, who feel more self-worth if they are part of the customer service. Sometimes locating the Helpdesk within Reception is a good idea. Don't shy away from creating an image for your department. This can help to give credence to the service and professionalise an area which is sometimes seen as a peripheral service. Your image should be dynamic, because the service you provide is dynamic and responsive. A front-of-house location puts your service on the map and shows all other departments that you are professional, self-confident, vibrant and dynamic.

> **Note to self**
>
> The Helpdesk needs to be dynamic and customer-oriented at all times, a showcase for the professionalism of the department.

Your Helpdesk system should be:

- capable of logging and tracking jobs
- capable of setting parameters for completion, e.g. within 24 hours or five days

- capable of recognising and dealing with "planned" works, while still using the same credentials of checking completion within an allotted time
- capable of storing comments so that you can update staff on progress
- capable of storing building-specific or even department-specific data
- a channel for issuing work

Sometimes it may not be practical or financially viable to have a computerised package for issuing and tracking jobs. Realistically, for a Helpdesk work efficiently, it needs to be computer-based, even if that is simply a spreadsheet which details jobs. You need to be able to follow and track jobs to make sure they are completed. It is also far easier to produce statistical information using a spreadsheet.

Customer service department

Today's sophisticated society is more than capable of making judgments on what is good and what is bad customer service. You need to be clear to your Helpdesk team and, indeed, all your staff, that customer service, or from the students' perspective in an educational environment, "the learner journey," is of paramount concern and time spent emphasising the message about customer service will be time well spent.

Audit department

Your Helpdesk should also be the arbiter of poor service. There should be supervisory staff in each department checking work and making sure you meet your targets on quality, but this is too good a chance to miss. You have the perfect opportunity to improve the quality of your service by getting your Helpdesk staff to contribute to your audit service by:

- undertaking quality audits on services
- undertaking customer satisfaction surveys on the quality of the service and works provided
- providing statistical data on quality to report to senior staff

The ability to self-analyse performance and learn and adapt to the changing face of the site are key to the success of your department. These data are essential in achieving this. Your Helpdesk should be developed to provide data to you in a format you can use effectively and in a way that can:

- justify your department in terms of value
- show the relationship between your workload and staffing levels
- allow you to bench-mark your service
- provide credible data to show your performance in relation to your peers in other institutions
- allow you to set realistic goals and targets
- demonstrate your achievement of targets set and complied with
- show risk where appropriate

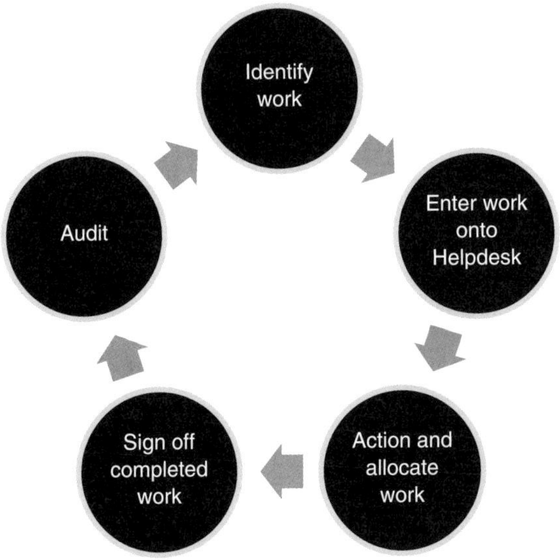

The cycle of work through a Helpdesk

Helpdesk software specification

Your computerised system should be a database of your entire building stock, providing you with a breakdown of your site in terms of what is in each building and allowing you to access more detailed information at floor level, including the size and type of space and rooms available with the appropriate occupancy level. More sophisticated systems also allow CAD plans to be uploaded to show your staff the locations of rooms. This can be invaluable for staff who do not know your site well or who are off site. Further levels of detail can show M & E assets in any area and allow special requirements such as permits to work to be uploaded. Additional information such as guidance notes can also be provided.

Ideally the system should be able to:

- interface with Outlook to use the email facility
- email jobs to the appropriate team member
- cut and paste information into and out of the job sheet
- move responsive jobs into the planned category if needed
- have a section for room schedules, e.g. availability of the room where the repair is needed
- add follow up jobs, e.g. paint after a repair
- show the top ten most fre-quent jobs based on performance indicators across the bottom of the screen

Note to self

Set up the Helpdesk to be critical of the service it provides.

System requirements

The following are recommended:

- the capability to upload full details of each space on the site as a spreadsheet
- a "Go to" facility to access a predetermined job reference, the last place you looked at, or the top or bottom of a list
- the ability to print both single and batch jobs
- the ability to reprint orders without duplication
- the ability to allow individual users to customise their front-end and to save settings
- online tracking of jobs by the end-user

The system should also be capable of holding data on all your contractors, such as:

- contractor's name
- contact name and number, including email address

- details of public and employer liability insurance
- compliance with health and safety policies
- permit to work requirements, such as working at height or hot work permits
- an audit facility, so contractors can be tracked on the quality of their service
- the ability to automatically print vehicle and visitor passes
- the ability to generate an email to the PDA of the responsible manager when the contractor signs in

The system should also be capable of uploading all your PPM jobs. There are two types of PPM: internal and external. External PPM jobs are those covered by external contractors, such as lift maintenance, boiler maintenance, fire and emergency lighting and fire-fighting equipment. Your system should be capable of:

- logging contractor activity on a calendar
- sending a reminder email to the contractor
- providing a log in–log out facility to give time and date stamps
- providing details of work required
- allowing a job to be signed off by the maintenance supervisor via PDA
- providing information to the contractor, e.g. permit to work requirements

Internal PPMs include the cyclical and maintenance jobs undertaken by in-house staff including: cleaning of gutters, blowing down calorifiers, running underused taps, *Legionella* testing, oiling door closers and hinges, changing filters and rotating pumps. Your system should be capable of the following:

- sending an email to the maintenance or site supervisory team (person-specific)
- sending an email to each team, e.g. morning and afternoon shift supervisors
- providing details of works required
- interfacing with stores and a basic list of spares
- signing off a job by supervisor or manager

- providing permit to work requirements
- giving online access to risk assessments
- giving access to guidance notes on PPM processes
- interfacing with staff holiday dates ensure that staff are available

A Helpdesk strategy is given in the Appendices.

Post team

A thriving organisation will have a busy post room, with vast amounts of mail, both inbound and outbound, on a daily basis. Factor into this the amount of internal mail and documents that need to be dealt with on a daily basis, together with deliveries from suppliers, and you could have a logistical nightmare. Post rooms may have a steady and even workflow, but from time to time will explode in terms of volume.

Your structure needs to be able to cope with this ebb and flow of work and be able to manage peaks and troughs seamlessly to ensure a continuity of workload. It is no use pushing your staff to breaking point during times of high workload, then allowing them to be idle when the work dries up. It is much better to even out the work or spread it between different functions. With that in mind, it is useful to overlap services to allow peaks and troughs to be evened out. The most obvious way is to amalgamate Post with the Helpdesk. This way you have the opportunity to increase the skills of staff from both disciplines, while at the same time creating continuity and allowing succession within the department.

Don't underestimate the amount of space you need to undertake this function, not just safely, but in an environment which is conducive to work. Create space for your franking machines and folders and try to create work flows which require the least movement between functions. Allowing room for post to be accurately sorted into batched sizes will speed up the franking process.

Don't forget this is a service in which you meet customers face-to-face. Staff will visit the area each day to pick up or drop off mail. Making the whole process look professional will improve your image within the organisation.

Deliveries out to departments

There are various options for the delivery of post out to departments and you really have two considerations here: spread of locations and available manpower.

You may be lucky enough to have a single site where all mail can be delivered to a central point and then distributed around the site, or you may have a central base with a number of off-site centres miles away from the main hub. Your considerations may therefore be about creating various mini post centres or arranging a central pick-up point for all your post to come back to your central sorting office. There is no hard and fast blueprint for this, so this is one area you need to look at carefully and weigh up the options available. Rather than delivering mail to multiple areas, why not look at a single drop in, say, the administration area of each centre, allowing other staff to continue with the internal distribution of post from that point onwards. Considering all your options could provide you with a solution that has not been considered before.

You may be able to distribute parcels and post directly to work areas from your post room as they arrive on site. This shows good customer service, but is extremely costly and inefficient. The best and most efficient way to deliver post is once per day. This is best timed with the early delivery of the post to site, so that all departments receive the day's mail before a set time, say 10 a.m. Any residual mail can be tied in with the delivery of reprographics materials, parts or documents later in the day by your site supervision team. This frees up your Post staff to undertake Helpdesk duties; in addition, early post sorting can be undertaken with the assistance of the Helpdesk staff to share workload and encourage teamwork and bonding.

Be careful about requesting timed delivery from the Royal Mail as this will increase your post bill considerably. Currently standing at nearly £4000, the cost of a timed delivery can sometimes be seen as a luxury. Ask your departments if there is a special reason for having the post early. You may find only one department requires an early post. Look carefully at timed delivery as removing this can be a quick way to save money.

Another way to save on your postage costs is to batch your mail in sizes and in large volumes. The Royal Mail allows this type of mail to be discounted if it is batched in groups of more than 1000 and the addresses are typed. This can create substantial savings on your post bill.

Parcels and deliveries

The controlled delivery of goods to site is part of the general workload of this department. Ensuring and logging the delivery of parcels and supplies and how those deliveries are made is a crucial role in the control of stock and financial management of the busy post room.

It is important that all deliveries come to a central area whenever possible. This allows you to control what has come on to the site and when it arrived.

Deliveries should be signed for and logged to allow tracking of those goods around the site. It is easy to lose control of valuable stock coming on site when you don't know when it came, who received it, who moved it on to the end-user and who accepted responsibility for it when it finally reached its destination.

It is good practice to keep a tally of all deliveries and post coming on to and going off the site. This will allow you to undertake data analysis and assist you in justifying changes in the department as a result of changes in workflow. It may also help you to make the case for additional staff when workloads increase.

Note to self

Start to pull together statistics on how much mail is coming into and going out of the post room. This will help with budgeting and staff resourcing. Ask staff to help collate the figures.

Cleaning team

You can never win with this service, as you will rarely be able to provide perfection. Cleaning, when done well, is rarely noticed. It simply blends into the background and is taken for granted. However, as soon as it goes wrong, your credibility diminishes rapidly.

Cleaning is likely to be the largest single service that you provide and the largest single budget you handle. Everyone thinks that they are a cleaning expert and will only be too pleased to tell you how they would deal with cleaning, how they would improve cleaning and how they can save money for you. When cleaning is sourced in-house, it also likely to be your biggest drain in terms of the time you to devote to HR and supervision.

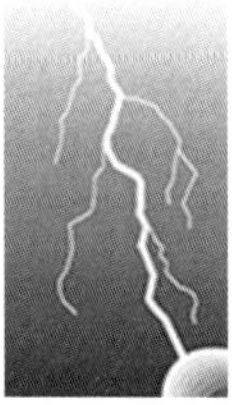

★★★★★★★★★★

Set up a focus group made up of staff from various departments to look at cleaning. Consider: times on site, standards, feedback and service improvements. Drive the service forward with high quality at all times and make your cleaning service your prime objective. Get this right and the rest will follow.

★★★★★★★★★★

Timing

Cleaning should never be seen. This is what is generally thought by most people. The cleaning process can get in the way of the day-to-day business of the establishment. No one wants to hear vacuum cleaners; the silence demanded by most office functions will generally preclude you from providing the service when the core activities of the establishment are taking place. Therefore your cleaning service is generally provided early in the morning or in the evening. This is the time when you are not on the premises yourself and therefore makes it difficult for you to bond with the service. It does not allow you the opportunity to interface with the staff who provide the service. If your cleaning staff don't know you and never see you, they may never understand exactly what you want from the service. As a manager, you need to be visible to all staff. You have to be approachable and you have to be seen to be human. You can never command respect from your staff if they don't know you. The way in which you supervise and manage this service therefore needs to be carefully mapped out.

Note to self

The cleaning service is a no-win service, but is probably your biggest budget area. Make sure you are clear about the service you want and give it the attention it deserves.

It's only cleaning! If you take that attitude, then shame on you. Your cleaning budget can be as much as a third of your entire budget. If you are prepared to under-manage this amount of budget, then you are asking for trouble. If you are seen as a manager who is only around when there are problems, this can further alienate your cleaning staff. If the only time you see them is when you are dealing with sickness or discipline, they will come to see you in a negative way.

There are times when you need to be available outside the core business hours. Your cleaning staff will work for you far better if they feel in touch with you. Try to arrange to be around on a regular basis, not every day, but perhaps once a week. Make sure you take time to walk around and have a chat with your cleaners. Block some time on your Outlook calendar and treat this time as you would any other meeting.

However, it has been proved that better productivities are achieved from cleaning staff when they are cleaning while other staff are working. If you can convince your senior managers of this, you will achieve better productivity. If your cleaning staff become part of the overall team, then they feel part of the whole process and an integral part of the FM department. Also, because the service is seen, other staff in the organisation will know that you are cleaning and you will receive far less complaints about the service. It is one of those feel-good factors; if people get to know the service they are far more tolerant of it.

Supervision

As you are unlikely to be around all the time, make sure that your supervision team is on your side. Make an extra effort to engage with them and make them feel part of the team. Take time out to have meetings with them, either formally or informally, and make sure they know what is expected of them. Create an SLA by which you can measure the service you provide and set them targets to achieve using the SMARTER target system explained earlier.

Warning
Never relax your control of a cleaning service. It needs constant management.

Your supervisors must be well trained. Your cleaning supervisors must be trained to the level where they can step up to management in just the same way as everyone else. You are training them to be your eyes and ears so that they can promote the organisation's aspirations. If you get this service right, the rest will be easy to conquer.

The "poisoned chalice"

Cleaning is a very difficult service to get right because it is usually unseen, so no one believes it takes place, or it just blends into the background. Staff rarely see it as a professional service as "anyone can clean." This is absolutely wrong! Anyone can clean, but not everyone can clean well. When you get it right, no one wants to know, but if it goes wrong, people cannot accept that you made a mistake with such a "simple" service. To understand the cleaning service is to understand its true complexity and the sheer effort needed to make it perform well.

Cleaners can be transitory. Some have stayed with the job for years and want nothing more, yet to others cleaning is seen as a stepping stone to a full-time job, based on the view that it is easier to get a job when you have a job. Sometimes cleaning can be a high turnover business, especially recruitment, training and the administration associated with HR. This can be very frustrating when trying to achieve stability in a department. Because of this transient nature, the overall application to the job can be less than from other departments. The motivation of cleaning staff can also be very difficult, so the more you engage with your cleaning staff, the more likely are likely to succeed.

Value added: sustainability and security

Your cleaning team should never be seen as just cleaners; look for value-added services that can be provided. Some managers miss some of the more obvious value-added services that can be provided by your cleaning staff in the execution of their duties, such as sustainability and security.

Sustainability

Not only can your cleaning team directly influence your sustainability performance, but there are successes that can be achieved by using your cleaning staff in a proactive way to improve your sustainability.

How often, if you are up early enough or behind late enough, do you find all the lights switched on, windows wide open and the heating on full? Encourage your cleaning staff to operate in one room at a time, turning lights on as they enter and off again as they leave. Most modern buildings have thermostatic valves on the radiators. Ask the cleaning staff to turn these down rather than open the windows. Cleaning is a physical job and working in a hot environment makes it more difficult to work, so cleaning staff may benefit from the heating being turned down or off early. However, don't just turn off the heating without discussing it first, as you may end up with a walk-out on your hands.

Ask your cleaning staff to log off PCs while they are working; it will more than pay for itself in terms of time from the savings you will make. Other items such as heaters, projectors, photocopiers and air conditioning can also be turned off early, saving further money. If cleaners are the last into the building and don't turn these items off, then they will be left on until the next day or over the weekend. In an educational environment, you could find that they are left switched on for the whole of the summer break. So a little bit of training in this area will save you a lot of money.

Be careful, however, what duties you load onto the cleaners. If you start to stray into other duties, you may find yourself faced with a fair and equal pay claim which you would find difficult to contest, e.g. if you ask cleaners to lock up and move things around, what is the difference in duties between your site supervisors and your cleaners? If your site supervisors are paid more for doing the same job, why should the cleaners not ask for the same pay for commensurate duties? So when mapping duties, make sure you can clearly justify the difference between one group and another.

Security

> **Note to self**
>
> Think laterally about what value-added services you can achieve from your cleaning service. Don't be afraid to think out of the box.

Your cleaning staff are sometimes used as security staff, your eyes and ears around the site. Train them to be vigilant about strangers in the buildings and encourage them to challenge, in a polite way, people who don't seem to be in the right place at the right time.

Ask them to work in tandem with your security and site supervision or caretaking teams. Asking them to lock as they go will reduce the locking down time of your caretaking team. Using them as a valuable resource for security will provide a far more robust, secure environment, which is also good for safeguarding issues.

Consumables

You will find a fixed link between the size of your cleaning service and the amount of consumables used. The more cleaners you have, the more cleaning materials are used and, consequently, the more you will spend on items such as cleaning chemicals, black bin bags and toilet rolls.

Because of the timing of the service, it is easy for these kinds of consumables to leave the site unchallenged. A few items leaving in this way throughout the year can make a huge difference to your annual stock levels. It is therefore necessary to have a robust stock control system to ensure and track your cleaning stock. Having a good stock control system is enough to deter minor pilfering. Remember that you need to monitor this service in the same way as every other.

Security team

Society has changed dramatically over the last 20 years. However, educational premises, in particular, have always been open and accessible, the main exception being junior and infant schools which have, for some time, been much tighter on security. Colleges are generally open, and because of this may be easy targets for theft and other crimes. The security of your premises is therefore one of balance.

Do you secure all entrances and exits and put guards on every door? This indicates to the outside world that you have a problem. Or do you go the other way and keep open access with a low key guarding structure, potentially putting yourself at risk of severe criticism if an incident results in media attention?

The basics of the security service are as follows.

- *Policies and procedures.* Link to your strategic objectives, one of which is likely to be safeguarding in an educational environment, and ensure your policies and procedures mirror each other.
- *Define your service and publish your outcomes.* Set your targets based on the statistical information you have collated and use these as bench-mark statistics to allow you to set further targets.
- *Parameters of the service.* Define exactly what the service consists of in terms of the following specific areas of concern: safeguarding, intruders, drugs and alcohol, CCTV and patrols.

Safeguarding

Curriculum departments deal with the pastoral side of safeguarding in educational institutions and your strategy should dovetail into that curriculum response. This is one area where you need to have a watertight connection between the service you provide and the service provided by curriculum departments. The two services must talk to each other in a coherent way to share intelligence that will provide that holistic service demanded by legislation. When setting out your meeting strategy, make sure that the pastoral staff receive your minutes or, better still, join the meeting and participate. This is the only real way of ensuring fusion between the two services. Don't forget that safeguarding and its principles are just as important to a Facilities Manager in a shopping centre or a manufacturing industry. Strands of legislation will apply to all areas of FM in a number of different environments.

Intruders

You need to protect the site, staff and, in an education setting, students from intruders. This may be difficult because of an organisation's open door policies with full accessibility. This is what puts colleges at the heart of the community. Over time, many routes through sites have become public rights of way and this makes it difficult to challenge intruders and strangers on site. The same principals of securing a site are important in all organisations and your security risk assessment will help to decide what level of security you need to implement.

However, you must seek to prevent intruders from entering the premises. This can be achieved in a number of ways, particularly if financial or political reasons prevent you from using swipe card access on every door. Simple measures, such as requiring all staff to carry and show ID cards, can be useful if implemented sympathetically. Incentives such as reduced or discounted rates in cafes will encourage their use in proactive and productive ways, rather than labouring the negative aspects of security. This can also create a bond between staff and the company by making staff feel part of the organisation. In this way, people who are not wearing their badges become instantly recognisable as strangers.

One recent trend is for security services to dress down to give a more relaxed feel to the site. This can work by preventing strangers from immediately spotting security officers. The negative side is that they may become bolder if they feel there is no security on site. There are pros and cons for both views and it is up to each individual institution to decide what is best for their site.

Drugs and alcohol

Drug and alcohol abuse is becoming far more prevalent in society and it is a difficult subject to deal with without appearing to admit that you have a drugs problem on site. You may choose to carry out drugs checks or searches, which can introduce their own problems. In an educational environment, targeting your new student cohort with information and advice on drugs and providing support, particularly at the beginning of their tenure and repeating that message as the term progresses, is a good method of tackling the problem. Make sure that this is a combined exercise between your department and the pastoral staff.

You will probably find that your statistics for drugs in particular peak during the September intake. You may also see a sharp increase in the number of people dealing drugs on site and therefore may need additional resources early in the curriculum year. Many dealers see the start of term as an incentive to influence the new cohort, generally 16 to19 year olds who have just left school and are easily persuaded into buying drugs to experiment. Your strategy must be one of prevention rather than cure. This is an area that is of utmost importance and one that colleges need to tackle early in the year.

In other environments, you could display signs on zero drug tolerance for visitors to site. This promotes the social corporate responsibility of an organisation in, say, a shopping centre. You could also target drugs as an issue through staff newsletters or during induction procedures.

Closed-circuit television

Closed-circuit television (CCTV) can be a very useful tool. Even when not in use, a well-designed CCTV system is a great deterrent to intruders and drug dealers if they think their actions will be caught on camera. To be effective, your system needs to be well planned and to give as full coverage as is practical so you are able to track people as they go around your site. There are basically two levels to be considered: internal and external.

Internal cameras can be useful, but can also be seen by staff as "Big Brother is watching you." Internal cameras will generally be fixed and focused on a particular area, such as a doorway into a building, a high risk area where money is stored, or food outlets where cash may be taken. Most modern systems will be digital.

Care must be taken when specifying your system. Make sure you are buying a good quality, tried and tested system. If necessary, speak to similar companies and see what has worked for them and what has not worked. Make sure you size your system to ensure you have enough continuous recording time. Also make sure that your storage space is large enough to allow expansion of your network. There is nothing worse than spending thousands of pounds on a system, only to find you only have one or two days of recorded storage space. Ask for guarantees of this from your installer. Try for five frames per second with 30 days of recording time, allowing for 25% spare capacity in the system for upgrades.

It may not always be possible to fully cover a site out of hours. Indeed, it may not even be necessary. If you have never had issues at night and you are confident that this will not change in the near future, then why invest in overnight monitoring of your CCTV systems? You can always review this to make sure that the situation has not changed.

Patrols

This is the aspect of the security guard service that most people identify with: your guards walking around the site keeping a check on things. It is far more than that, however. Your guarding service patrols are the visible face of the organisation. They need to be professional in their demeanour and to represent authority without being intrusive or threatening. They need to support staff when issues arise; visitors and guests as well as staff need to know that they are protected from external influences.

In an educational environment, the security patrols need to provide a friendly, but professional, service which engages with the students while making sure they know the parameters of acceptable behaviour. Groups of students will naturally push the boundaries of acceptable behaviour. Care must be taken to ensure that the guards are suitably trained. They must be routinely reminded that they should not become over-friendly with students or form liaisons with them, which are both unprofessional and inappropriate. This can easily create problems on site. A good team of security guards can provide you with much peace of mind about your service and a positive, friendly security team with good customer relations will have a positive effect on your business.

Structure your team's breaks around natural breaks in the core activities, making sure that the guards are not on their breaks when incidents are historically most likely. Times such coffee breaks and lunch times are times for the service to be high profile.

Your strategy on reporting should be carefully thought through before you begin. This is because, as with any other service, the information you are providing must be accurate. If you can get your reporting strategy right in the first place, then your data further down the line will be more reliable and accurate and, importantly, consistent.

Statistics

Statistics, if collected in a consistent way, can help you to focus your stretched resources into areas where they will be most effective. Your statistics can also help staff to focus on at-risk groups and target issues such as bullying and harassment.

Your statistics can cover:

- *Geography.* This helps you to identify hotspots on the site where most incidents happen, focusing you on where your patrols need to be targeted. It helps to put a visible face in an area where it will send out the best message to those who may cause mischief on your site.
- *Groups.* There may be issues with a certain age, gender or ethnicity group and targeted reports will help you to focus on these in particular.
- *Times.* There could be times of the day when the problem becomes more noticeable. Breaking your information on incidents down into times of the day will help to pinpoint times when the site is most likely to have problems. This can then help you to provide the right number of guards at the right time.

Security is clearly an area which can have a massive impact on your organisation in terms of community perception. Bad press in relation to issues surrounding security have the potential to cause a substantial threat to the organisation. The balance of the service provided needs to be constantly reviewed to ensure it fully meets the requirements of the organisation. Using risk assessment to look at security will be beneficial in showing where the issues are and will focus you on service provision and issues you need to target.

Aide Memoire: In-house teams

- Get to know your teams and learn what skill sets you have available in the individual teams.
- Ensure you are able to quantify all the work undertaken by your individual teams.
- Check the performance of individual teams and ensure consistency of performance across all teams.
- Treat all teams the same and never allow one team to become a favourite.
- Ensure you can track how much each service costs by developing individual cost centres.
- Ensure the work of your teams is documented in terms of levels of service and content of workload.
- Ensure the individual services are being provided when required, i.e. matched to operational hours.
- Make sure each team's work can dovetail into the next team and encourage crossovers between departments.
- Try to promote a friendly customer-oriented service.
- Ensure your Helpdesk sits at the heart of the service you provide and becomes the hub and driving force of the service.
- Ensure services such as cleaning become more visible.
- Ensure you don't miss the potential for using your cleaning service for sustainability issues such as turning thermostats down.
- Make sure your security services dovetail into your other services.

Chapter 7
Maintenance including new builds and space management

Maintenance procedures are carried out when something needs corrective or pre-emptive action to be able to function correctly. Good maintenance helps to provide optimum performance and longevity of your buildings and plant.

When you arrive in your new post, you will probably see that a seemingly disproportionate amount of money is budgeted for reactive maintenance. If this is the case, then your FM department has probably been poorly run in the past. You now have the opportunity to develop your maintenance strategy proactively in a way that shows you are a credible Facilities Manager. This seems to imply that your predecessor has applied his or her resources poorly; however, you will generally find that many departments "fire-fight" maintenance and never get to grips with the philosophy of planned maintenance.

Types of maintenance

There are three distinct areas in maintenance: reactive, planned and planned preventative maintenance (PPM). Most people will say that PPM and planned maintenance are one and the same. In my opinion, they are not, as explained in this chapter.

Warning
Never be tempted to buy cheaply, as inevitably you will need to buy the item again in the near future.

Reactive maintenance

This is simply repairing things as they break. Reactive maintenance can be one of the biggest drains on your budget. If things break, it can be for a number of reasons.

The first reason is the poor quality of equipment or finished works. If you buy or repair things in an inferior way, then they are going to break more frequently. There is much to be said for the old adage "buy cheap, buy twice." When we are striving to drive down prices, there is always the temptation to reduce specification. If we are rushed in our decisions when procuring, we can fall into the trap of not writing or choosing specifications that give longevity of the finished product or work. Remember always that you need to show due diligence when purchasing equipment and services for your organisation. Due diligence not only means being studious about the way in which you procure your goods and services, but is also about paying attention to quality to ensure you protect your organisation. Your mantra should be: quality, quality and quality. If you are not buying, procuring and managing your purchases based on quality, then not only will you make more work for yourself in the long term, but you will also be labelled as someone who does not care about the service you provide.

Another reason for breakages is poor maintenance as a result of ignorance. There is no point in spending hundreds and thousands, if not millions, of pounds on new buildings or refurbishments only to focus on something else and forget to maintain these properly. You will often see this where two or three years down the line a fan burns out because it has not been serviced. Kitchen fans become clogged up with grease because no one has put in place a regime for changing the filters. This, in my opinion, is ignorance of the systems and a general lack of care; due diligence again.

Poor maintenance can also be a result of budget pressures. This is potentially the worst position for a practising manager to be put in. Sometimes we all have to bear the burden of a reduction in budgets, but this must be done in a way that reduces the long-term liability of the services you provide. Risk assessment must be used to understand the long-term liabilities of not maintaining a piece of equipment, or dispensing with a service provider just to save money. In the long term you could potentially be creating a far bigger problem by not maintaining a piece of equipment. You need to understand this long-term view before committing yourself to a course of action. You cannot cut a

Note to self

When you obtain a new building, make a list of all the plant, equipment and structures that will need maintaining. Put them all into you PPM schedules and start monitoring, managing and maintaining from day one. It will almost certainly help you to identify maintenance needs early, and will also validate your defects period if everything is fully maintained.

service or maintenance regime without ensuring your senior managers, directors and governors also understand the long-term implications of doing so.

Poor skill levels of operatives or technicians can also result in higher maintenance costs. If they are not skilled to appropriate levels to manage and maintain equipment and processes, it follows that you will need greater resources to put things right when equipment fails. It pays in the long term to increase the skills of your staff. This will also reduce the frustration you can feel when seeing equipment continually breaking down as a result of your operatives not using it or maintaining it adequately.

When staff are under-resourced, they are also under pressure to do things quickly. This can result in jobs not being completed properly. For example, you may have a heating problem. Your heating engineer or plumber cannot immediately find the problem, but restarts the plant and the heating comes back on. They then go on to do other jobs and forget to return to resolve the initial problem. It could be days, weeks or months before the problem is remembered and resolved. This can lead to other equipment around the site being turned off or disconnected, preventing the system from running effectively.

Planned maintenance

Planned maintenance is somewhere between PPM and capital replacement. Most people see capital as items of equipment or additions/refurbishments that add value to the property and are therefore capitalised or written down in terms of value at the end of the financial year. This is true to some degree, but there are many minor replacements in a system that are not of sufficient value to merit capitalisation, such as pumps or drives. Sometimes these items are classified as minor capital.

Planned maintenance means replacements, upgrades or refurbishment that need to be planned to be replaced ahead of obsolescence. They are not large enough to be capitalised or to form part of a maintenance regime or PPM. Clearly, replacing pumps before they break is not maintenance, but replacement of minor capital. Rarely, however, will you see any manager replacing things before they break, and certainly not if, through good maintenance, the item continues to work at its optimum performance. It is, however, important that a balance is found to provide and account for the potential to replace, even if you can avoid it. This is generally done by creating a "sinking fund," which can drip-feed a fund to allow the replacement of equipment before it fails. This is more of an art than a science, but a good Facilities Manager, with a number of checks and balances, can get it right most of the time.

What is a sinking fund? On a very simplistic level, a sinking fund is a savings account that you regularly put money into in predefined amounts to pay for something you wish to replace. However, this invariably becomes more complex. Rarely is a regular deposit into your account a random amount; it is made up from complex depreciation rates. Depreciation is a financial measure by which you write down the value of an asset over a period of time to accurately measure the value of your estate.

It is not practical to continually feed into a sinking fund on the off-chance that you will need it. This is not good practice and does not make best use of the resources of the institution. Your sinking fund should be a predictive fund for future pre-failure replacements, while at the same time showing sensitivity to budgets, which will please your finance department. It is always a good idea to keep on the good side of the finance department, as there are times when you will not always be popular, particularly if you get this predictive element wrong and need ask for more funding.

You should place a relatively similar amount of money in your sinking fund each year. This ensures that it builds each year and allows you to spread your cash flow evenly over a period of years. You need to be able to show that you are managing this fund correctly and that the amount going out each year approximately matches the amount going in.

Planned preventative maintenance

Planned preventative maintenance covers tasks that are carried out at a predetermined frequency to maintain the efficiency of a piece of equipment or a process. You carry out PPM to stabilise problems or, preferably, to prevent problems occurring in the first place. When budgets are cut, PPM is often one of the first things to be forgotten or axed.

Organisations sometimes get into the mind-fix of waiting until something breaks before mending it. This can be designed out using your internal processes and practices. Using quality audits can identify when things are starting to go wrong and this can create ownership. Without ownership, no one will care if things are not done and your staff will walk past things rather than repair them. Because they don't have ownership, they don't have pride. You should develop the culture of room reports, where items are checked and staff held accountable if something is found to be defective. This soon begins to build pride and, in the long term, will save money spent on minor repairs. Your plumbing or heating engineers can carry out the equivalent of a room report in plant rooms by monitoring conditions and creating ownership.

External PPM is undertaken by an external person or company; in other words, it is outsourced. These are specialist jobs which require specific skills. These jobs are generally outsourced because they are uneconomic to undertake internally. As a department, you cannot afford to recruit staff just to deal with a specific area unless you have sufficient volume of work to ensure continuity and keep your staff fully engaged. This kind of PPM is more difficult to monitor as you can never be sure the contractor is complying with requirements. Simple follow-up checks after service can help, such as writing the date they were changed on filters.

Internal PPM covers all the regular maintenance undertaken by your own staff. This list will be ever-changing as the type of equipment and buildings on your site change. This is another area that can suffer from service drift if the service changes and the list of regular PPM is not updated.

Your PPM should link into your QMS. Linking to quality gives you the ability to review and audit your PPM, which, if done correctly, will identify any gaps and provide you with a methodology for updating your list of PPM.

How to establish PPM schedules

You will need a list of all the processes and procedures that will eventually form your PPM list. You can extract these from a number of sources.

If your Helpdesk is set up correctly, then information extracted from its records and statistics can be a massive aid. Look for patterns. Are there any jobs that are carried out on a regular basis? Is there something that is getting put onto the Helpdesk at the same time every month? If so, why are the Helpdesk team repeatedly entering them? This kind of job should be set up and carried out automatically at prescribed intervals. It is then one job less for your Helpdesk team to put on the system. Most Helpdesk packages will have a PPM module, so make use of it!

Your O & M manual should, if set up correctly, provide you with all the information you need about the building and what is contained in it. Extracting all the maintenance jobs from this document should be relatively easy, but you will need to look closely as sometimes these jobs are embedded within other product and machinery information. One way of doing this is to get your build team to provide you with all the maintenance processes as a spreadsheet. This can be set out so you can upload it directly to your Helpdesk system.

Your staff will be able to tell you about all the jobs that they do on a regular basis and the essential information you are likely to need. This will also engender buy-in by staff, creating that all-important ownership. Staff can provide information about the following aspects.

- *What.* What are the processes and are there any linked or similar processes that can be undertaken at the same time?

- *Where.* Your staff can tell you exactly where a piece of equipment is. Sometimes, even with theoretically up-to-date drawings, it can be difficult to identify where pieces of equipment are located and, if you don't document these items and someone leaves, you might find it difficult to track them down. Typically these are items such as ducting dampers, three-port valves, drain-offs, BMS controllers and sensors.
- *When.* When is the best time of day to undertake the task to avoid disruption? Sometimes you can be frustrated by a PPM being on the system and your staff never being able to access the job because it is in an office where workers cannot be disrupted. It is far easier to work your PPM to suit occupation patterns.
- *How.* Staff can provide details about how they do the job and give a logical process for you to write down as a procedure.

Note to self

Make sure you have a formal specification of works for every PPM item that you have on the system. That way, you can be sure that they are being undertaken in a consistent way.

Note to self

Borrow ideas from other Facilities Managers; don't continually reinvent the wheel.

You can also use your peers to give you information. Most Facilities Managers will deal with the same processes and do not mind sharing best practices. Use your networking skills to provide you with a ready-made set of PPMs and be prepared to return the favours.

Your PPM can serve purposes other than just maintaining your building stock. You can also use PPM to prove what you do as a department. As a department, you need to be able to justify yourself; monitored and managed PPM can emphasis the volume of work your department undertakes. Rigidly followed PPM can prevent litigation in some areas, e.g. staff reporting headaches and blaming dirty filters on air-conditioning units. An enforced PPM that changes the filters on a regular basis and documents when they are changed will save you many hours trying to track down this information. Likewise, if, as part of your PPM, you have a room check, you are able to document when furniture in a particular room was checked, so if someone traps their finger or cuts themselves you can prove due diligence in having checked the furniture on a regular basis. This again saves you issues with insurance claims.

Your PPM can have a huge influence on your budgets. If you can incorporate and dovetail your planned, minor capital and PPM functions together and apply your PPM processes diligently, you will reduce the amount of breakages and repairs you need to undertake and will move the balance of maintenance from responsive to planned.

PPM specification

It is important to put specifications behind each of these PPM processes. No matter how simple the PPM is, e.g. cleaning a foyer window, the work will be undertaken differently by each person who carries it out. This does not really make a great difference when the PPM is something so simple. However, if the PPM is to service aluminium window ironmongery and you want the hinges, handle and locking system oiled, it could be that one or more of the elements is missed. It is easier to monitor PPM if you can compare the works against a specification. If you have a computerised Helpdesk system, your instructions can be digitised and included in any work order.

Planned programme and planned survey

You need to know your on-site liabilities and what to plan for in the future. You also need to be able to provide quality information to justify your statements. The only sure way of achieving this is to have a planned programme and a planned survey of your building stock.

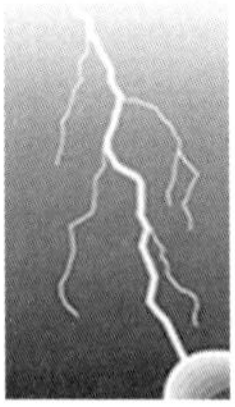

★★★★★★★★★★

Be sure that you deal with your survey in a consistent way. Draw up a template that fits all your requirements and use the template on each building in the same way. This will give a consistent feel to your survey results.

★★★★★★★★★★

What is a planned survey document?

This is a risk assessment of your building stock containing detailed information about your stock, its current condition, any failed or failing items relating to that stock, the anticipated lifespan of major elements and costs for rectification of failing stock.

Details of the building stock

A building by building overview of your stock, set out in a consistent way, helps you to access information quickly. You can then compare this with other parts of the building stock. There are no hard and fast rules about how this should be set out. The easiest way is to devise a system that works for you, or borrow one from a fellow Facilities Manager.

In general, the details of the building stock will contain the following information: basic details about the stock, such as size and age; ownership and legal responsibility, such as rights of way, tenancies and repair obligations; an overview or statement about issues relating to the building stock, such as dry rot; the long-term objectives for the buildings, e.g. are you planning to replace buildings or are you planning a major refurbishment in the future?

> **Note to self**
>
> Don't forget to include all your properties in your planned survey. Also include all your leased properties, but be careful to read the lease to determine what your repair obligations are. Only include these items in your survey.

Details of current condition

This is easier if each building is broken down into its constituent parts in a consistent way across the site. Your breakdown could include the following.

- Roofs (pitched): copings; covering type, e.g. natural or concrete slate; lead work, valleys and soakers; parapet walls; ridge tiles and hips; sarking; structural timberwork; and insulation.
- Roofs (flat): covering type, e.g. built-up felt, Kingspan; abutments, laps and joints; breathers and other penetration details; drains, internal gutters and downspouts; structural slabs; and insulation.

Details of any failed or failing items relating to that stock

For each of the elements of the building, it is then necessary to collect the following information.

- When the element was last replaced. This will give you some idea about when it will need to be replaced again.
- What condition is the element in now? Do you understand how far into failure items are and do you know when a failing item will completely fail?

> **Warning**
>
> The use of consultants to undertake your survey can be expensive, but sometimes it is worth the extra money. Be aware, however, that you can sometimes get exactly the same service from a contractor at a fraction of the price.

- What interdependencies are there, e.g. if one element fails, is another likely to fail also?

Details of anticipated lifespan of major elements

It is also necessary to work out the anticipated lifespan of each of the elements of the building. This information can be obtained from a number of sources.

If service and repair reports are set up correctly, they should give details of the equipment and indicate when an item is coming to the end of its useful life. This will warn you about a potential failure or the fact that an item of equipment can no longer be serviced as parts are no longer available. It makes sense to build this review into all your new contracts, so that in the first year of service you receive a full and detailed report on the condition and potential failure date of this equipment. This is particularly important where your item of equipment is specialised, making it difficult for you to make those technical assessments yourself.

You are not a specialist in everything, so call on specialists to provide you with information on the condition of specialist areas. Even in non-specialist areas you can benefit from having a second pair of eyes. There are two ways of doing this.

- Commission someone to make a full report of your whole stock or elements of it. This is potentially very expensive and most organisations do not have the available cash for this kind of survey.
- Speak to repair companies, e.g. ask a roofer to make a condition report for a roof under the guise that you are going to repair it next year and need to know what is wrong with it. Promise to put the company on your tender list. A few companies will look on this as an opportunity to build a relationship with you for other potential works and may carry out the survey for no fee. Caution is needed with this approach so that the contractor does not overdo their specification.

Manufacturer's books or O & M manuals can also help with this issue. When a new building is built, the O & M manuals will generally contain manufacturer's details about every element of the construction, including manufacturer's literature and service/maintenance recommendations for all the equipment that has been installed in the building. This information will usually include the expected lifespan of the equipment or any parts which may fail first. This will give you valuable data to put into your survey report.

Details of rectification costs

Your survey is of no use if it does not contain the likely replacement costs of failed parts. This information needs to be detailed and consistent to be of use. There is no point pricing or estimating half your repairs but not the other half.

Your senior management will be relying on your judgement to develop the risk register for the organisation. Your reports on the condition of the stock are a fundamental tool on which they can base decisions on the long-term viability of the site and its buildings. If your predictions are wrong, then you could end up maintaining buildings that would have been more economical to pull down and rebuild.

Planned programme

Your planned programme should be the culmination of all the hard work you have put into the planned survey and should give details of when you anticipate replacing major elements and target dates for doing this. Although your planned survey may deal with building elements as far as 10 or 15 years in the future, your planned programme will rarely consider more than the next two or three years. This is because the priorities of the organisation will change and you need to continually re-evaluate your planned survey details and update your risk register correspondingly.

However, simply identifying these building risks does not mean that there will be provision for you to deal with them at the predicted failure times. In the real world, things don't happen to plan. Your organisation's priorities will change, as will its direction. Funds may not always be made available, even though it would be financially sensible to do so. Your job is to continually re-evaluate and update the risk register. By all means point out the misgivings of underinvestment in your building stock, but never believe you have right to all the resources you request. This belief will always lead to disappointment and you need to be able to deal with this disappointment in a professional way.

Space management

Space management is one of the most difficult issues to handle. The estate is usually one of the biggest assets of an organisation. The way in which it is managed and allocated must be controlled and, unless you have infinite resources, it needs to be divided between its occupants in a logical way. Estates cannot be developed and expanded without some form of control: this is space management.

An institution will usually have a policy about how space is allocated. In an educational environment, where the allocation of space and requests from individual curriculum centres can be dogged by internal politics, it is essential that some guidelines are provided so that sound judgements can be made about allocation. Curriculum centres should be viewed as tenants of the buildings, with the college or university as the landlord. Allocation and residency must never be seen as permanent, but merely a commitment based on economic viability; that is, there is no use providing space to a curriculum centre based on current vogue subjects, only to find with a change in government policy that particular subject is no longer required. Space must be seen as temporary and based on sound economic reasons for continued usage. Space allocation must also be closely linked to your organisation's strategic plan.

A useful way to ensure that departments use the space provided is to charge for it. Departments are less likely to hold onto underused space if it costs them money. Space allocation should be based on a set of fixed parameters, e.g. square metres per student or member of staff, and those figures should be clearly published so that everyone can see that a fair and equitable system of allocation is in place. Educational funders such as the Learning Skills Council and Higher Education Funding Council for England have set space allocation matrices, which are generally linked to capital funding. Space should be seen as finite and department heads should realise that to create space, other space generally needs to be released to accommodate it. Organisations need to understand that the allocation of space needs to be monitored to ensure its effectiveness. This is known as a utilisation survey and it provides a notional value for how much a particular space is used. It is up to individual organisations to set an acceptable level of use and this is generally expressed as a percentage.

The allocation of space should be seen as a centralised function, with responsibility for allocation being taken by one individual or one single department. This task often falls to the FM department. Space should be allocated and monitored for all areas, including office and curriculum space. The use and allocation of space must be planned; unplanned moves or allocations should be avoided whenever possible.

Allocation must be based on sound financial or strategic reasons and not whims. Space, in educational environments in particular, is very costly and to have under-utilised space contributes nothing to the organisation and costs money best used for other resources. It is important when developing areas that space is identified and released before activities start. If this happens the other way around, your department could be criticised for its allocation of space. When developing space, never assume that this will be the final layout. Plan to be able to adapt by designing your space to be flexible in its layout.

In retail environments, empty space is lost revenue, so the focus moves from ensuring space is charged for to ensuring that space is occupied. Getting property and space back into service is far more critical in a retail environment. The level of re-letting or "churn" can have a dramatic effect on the core business, as empty retail properties are lost revenue. In a Mall, empty units make letting other empty space more difficult and this can become a vicious circle. Dealing with issues quickly in a retail environment is therefore paramount.

New builds and maintenance

You may be lucky enough to be constructing one building after another and thus be able to develop a genuine flair for project management. In this case, your estate may consist of perfect examples of newly constructed buildings; if so, you have been very lucky indeed!

Generally a new build is fraught with problems from start to finish. Sometimes your problems with a new building can last for years, giving you many sleepless nights. Very rarely will you end up with a defect-free building, and even with the benefit of hindsight problems can still occur. If you are aiming for a zero-defect building, this is just not possible! Your client's agent must be chosen carefully. You need someone who has specific experience of your sector and who has been recommended by other similar institutions.

Problems will often occur because of the speed in which building technology is changing. You will often include new technologies within your design and new and sometimes untried materials for fixtures and fittings. Don't be afraid of using new technologies, but be aware they may present problems in the future.

As the need for sustainability considerations increases, you may be asked to look at renewable technologies in your new building stock. You need to embrace this fully, making sure that you future-proof your organisation's power needs by reducing its reliance on electricity from the National Grid. Sustainable technologies such as ground-source and air-source heat pumps, photovoltaic and solar thermal systems, grey water systems, inverters, natural or passive ventilation and voltage optimisation should be considered.

Pre-build

You need to create a good working relationship with your design team. In today's climate, most projects are invariably run on a design and build

Note to self

Defect-free buildings are impossible to achieve. All you can do is to mitigate your risks by covering as many of the potential problems as possible.

principle. This can work well, but equally can be a nightmare in controlling the quality of all the elements within the building. Your contractor needs to be carefully chosen; insist on a contractor/ builder with experience in your sector. This can be filtered out at the PQQ stage. Equally, your client's agent needs to be completely up to speed with the requirements of your type of organisation. With a new build, you only have one opportunity to get it right and therefore all your actions need to be measured and thought through. It is far easier and cheaper to think things through before construction to than wait until the build has started and then try to deal with problems that should have been included in the original tender documents as client requirements.

Quality assurance (QA) is one area that most building companies will have considered carefully and should be very good at. The companies that I have had the pleasure of working with have had excellent QA procedures. However, I have never had a zero-defect building and I have never been completely satisfied with the quality of the product handed over. The builder has no long-term view of how you want to run the building, unless he or she has a vested long-term interest such as with a PFI project. Therefore QA on a new build only really relates to how the contractor views the finished product, not to how it is likely to be used, and there is a massive gulf between these two views.

Theory and information slot

A Pre-Qualification Questionnaire (PQQ) is a methodology of establishing whether a series of contractors/suppliers can achieve a set level of service level or compliance. It is commonly used by procurement sections of the company as a filter to cut down a long list of suppliers/contractors to a more manageable size, knowing after the PQQ that they all achieve a specific level of service or compliance.

Heating

Heating is an area that invariably creates problems for the FM team at handover. Unless you have been lucky enough to be completely incorporated into the design team and have sufficient skills to judge what is good or bad M & E, you are likely to have problems. Unless you have an M & E bias, you are likely to struggle in this area. Consider the following issues.

- Has the system been designed with the appropriate number of zones for heating? Beware of single-pipe systems over which you have no local control.
- Has the system been balanced properly and are you achieving consistent temperatures on each floor?
- Are there any rooms at the end of a run that are cooler than others on the floor?
- Does the BMS control temperatures at an individual level, or does it aggregate temperatures and give an average? The danger with the latter is that you can have very different temperatures on different floors, but the aggregation does not alter the final temperature. This leads to complaints from users at either end of the averaged areas; one area may be extremely hot and one too cool. This is particularly noticeable when the system is not balanced properly.
- Is the external temperature sensor in the wrong place, either in full sun or in a draughty position?
- Are the internal sensors sited wrongly, either near a heat source or in an inappropriate position? A classic case is where sensors are placed in offices. Individuals come in on particularly cold mornings and put local heaters on. This fools the sensor into thinking that it has reached the design temperature and it shuts down the entire floor, leading to complaints.
- Are the inverters and variable speed drives working? You will often see a variable speed drive or an inverter running at 100%. This is wasteful and, in fact, costs 3% more in electricity usage.
- Are all your external pipes lagged effectively against extreme weather? Most lagging is only designed down to minus 6°C and this can put your system in danger during extreme conditions.
- Is your lagging exterior grade? Interior grade insulation is often used and this starts to degrade after a short period of time, usually within two years.
- Is your external lagging protected? If not, birds will peck at it and then your rooftop pipe work will freeze and burst during the first exceptionally cold winter.

Ventilation and chilling/cooling

Ventilation is extremely difficult to get right in a new build as a result of high expectations. People expect their new building to be the most controlled environment possible. They expect it to be cool when they want to be cool and to have fresh air like a summer's day in the countryside. The reality is that today's modern buildings are so well insulated that marginal increases in temperature can affect the internal environment. New buildings can overheat very quickly and sometimes it is difficult to remove this heat from the building. When you have a mixture of ventilation types and cooling, people always want what their neighbour has. If they leave their own warm environment (say, 22°C) and enter a chilled area, they immediately feel that their own area was hotter than it should be. Ventilation is often linked to health and safety, so it is one area that you need to get right.

You should consider the following points.

- *Blocked filters.* Even marginal blockages on filters can reduce their effectiveness and, if you are in an inner city environment, the chances are you will frequently need to replace filters. Make sure you have a contract in place as soon as possible, preferably well before the end of the defects period. In this way your service contractor may be able to spot any issues for you to raise as a defect before the contractor hands over responsibility.
- *Changes to layout.* Many designs, particularly office designs, cater for a specific number of occupants and computers and the ventilation system is sized and designed to cope with these numbers. Moving additional people or computers into such a space skews the design parameters. What was an acceptable volume of air can be changed drastically and detrimentally by changing occupancy rates.
- *Blocked ducting or fallen fire dampers.* Sometimes fire dampers within air ducts are linked into the fire alarm system and, when operated, will drop to prevent smoke and combustion products moving through the system. If they are not automated, they will stay in the dropped position and your ventilated area suddenly becomes hot and stuffy.
- *Complaints about headaches or feeling tired.* This is an indication that your ventilation system is not working effectively. This is when you need to act, as inactivity could lead the way to claims that you did not respond to a health and safety issue.

- *Over-use of VFR units.* VRF units are combined heating and chilling units and these can be abused by staff. They are often left switched on, even when no one is in the room. Sometimes it is possible to limit the range at which the units can be used, e.g. setting a band or range of temperatures within which the user can alter the controls. This helps to reduce over-use and keeps your electricity costs down. Sometimes staff come in early on a very cold morning and switch all the VRF units on to full heat to take the chill off the room. This can affect the sensors for the BMS and turn off other areas, leading to complaints from staff. Education of users is the best way to get around this issue.

Electricity use and new technologies

One of the many justifications for rebuilding is the need to reduce the consumption of gas and electricity. Many consultants will extol the virtues of moving from an old building to a new low-carbon building. However, usually the opposite is true. Your old building will no doubt be a simple construction with plenty of ventilation. Many organisations have old Victorian or mid-1960s buildings, which are either of high thermal value or are simple tower blocks. Moving to a new building generally means you move from a simple building, with little or no technology, to a highly sophisticated building bristling with technology. There is always a shock when your utility bills, especially your electricity bills, are doubled. Again, this should be dealt with during commissioning, but rarely will you find that your building is operating to full capacity and efficiency. You will find many things running when they should not be. Most new builds now incorporate automated meter reading into the BMS package. Check your consumption over the weekend or when no one is in the building and you will probably find that this is far more than it should be, indicating that some electrical items are still on when they should not be.

New technologies are great, but you have to understand that until they become mainstream and tried and trusted, you may have problems with them. You may be introducing new maintenance regimes that have never been done before, so you need to factor in that things may not go smoothly.

Disputes

Be ready for problems with your new building. You may have to challenge things you are told and you need justifications for your queries. You will definitely be challenged back that you have not done certain things and have exacerbated problems, e.g. not changed filters at the right frequency, not tested and maintained inhibitor levels, or had the wrong settings on the BMS. Be ready for everything by ensuring that the recommendations in the O & M manuals are followed exactly, at least during the defects period, otherwise you will find your builder will have a get-out clause. However, not all builders are out to rip you off. A genuine working partnership where all parties are honest and open will provide a good new building that you will be proud of, but never assume it will happen without you working at it.

Aide Memoire: Maintenance

- Try to avoid buying cheap or inferior equipment: 'buy cheap, buy twice'.
- If you need to cut back on budgets, never arbitrarily cut back on PPM without making sure you tell your senior management about the likely outcomes in the form of a risk assessment.
- Make sure that your PPM is identified and put into place.
- Undertake planned surveys on all your buildings to ensure you have identified all the likely liabilities to the organisation.
- Ensure you work closely with your contractor and client's agent when embarking on new buildings.
- Check your buildings are commissioned correctly.

Chapter 8
Sustainability, use of resources and sustainable technologies

What is sustainability?

Sustainability is increasingly becoming an integral part of the FM role and most Sustainability Managers are now found in the FM or resources department. In a larger organisation the Sustainability Manager is often an entirely separate role because of the complex issues involved. So what is sustainability in the context of a busy FM department?

Sustainability means different things to different people and, as far as a busy FM department is concerned, it can become an integral part of everything that you do. If you are expected to take the lead on sustainability, you need to define what areas you are going to concentrate on, what specialists you need to bring in, and how you are going to report your progress back to the senior management team.

All FM departments buy in services for the site. Do we ever stop to think about the environmental impact of these services and supplies, e.g. the amount of carbon that has been used in their journey from source, through manufacture to your doorstep? Do we ever consider alternatives or even really consider sustainability as a real issue? The answer is probably no. As an active FM department, we swim in a sea of information and sometimes more information and responsibility seem too much to take on. Should

> **Note to self**
>
> Do not underestimate how much time you are likely to spend on sustainability. To do it justice, you need to carve out time to do it. Start by trying to get a regular time slot on your Outlook calendar, even if this is only an hour a week, and stick to it rigidly. It will pay dividends.

we bother with sustainability, or should we let others deal with this issue? In my opinion, we should definitely consider sustainability as part of our role.

The concept of sustainability has become increasingly important in recent years as we all try to reduce the resource intensity of our day-to-day lives. For hundreds of years we have been plundering the Earth's resources without much concern, but we now know that these resources are finite. Our growing population has experienced an explosion of consumerism and our planet's resources have dwindled to the point where, within a few years, we will face a catastrophic decline in the availability of fuel and raw materials. Previous generations have seen remarkable leaps in technology which have made us feel invincible. Many believe we will be able to develop alternative technologies to take the place of the traditional oil and gas deposits we have come to rely on. Nuclear energy sources, once seen as the answer to generating cheap electricity, have been called into question following natural disasters which have shown the vulnerability of this technology. The simple truth is that we cannot continue in this fashion without reducing our reliance on fossil fuels and giving greater consideration to how we can reduce the resource intensity of our actions.

How does this then affect individual organisations? Quite simply, we need to re-evaluate our systems to make sure we reduce resource consumption and, in particular, our use of carbon. This does take time and can be costly but, if undertaken correctly, this evaluation could bring the following benefits to your department.

- Reduce your carbon footprint.
- Raise the profile of the FM department.
- Save money, as waste is cash intensive.
- Give your organisation a marketing advantage over competitors.
- Show due diligence.
- Reduce your long-term reliance on fossil fuels.
- Increase your long-term robustness and reduce vulnerability with respect to utilities costs.
- Give an edge with key partners in terms of funding. An organisation which is seen to be managing its carbon use proactively shows that it is managing waste. When applying for grants or capital funding, a formulated carbon plan will give you the edge.

Sustainability in any business environment can be a journey of discovery, but also sometimes horror when you realise what you are using and what impact this is having on the environment. As with any journey, we must first establish our baseline so we can plot a course for the future. We must try to identify all the areas within the organisation where we use a resource and then calculate its impact on the organisation.

Improving the sustainability of your organisation

The logical stages to follow to increase the sustainability of your organisation are: identify all the resources used; convert these resources into a carbon figure; set a bench-mark figure; set future targets; and report progress and review.

Identification of resources used

This can be one of the most difficult areas to define. It will vary between organisations and may fall under the remit of more than one department. However, as one of the prime users of resources, the FM department should take the lead on this if there is no Sustainability Manager in post.

Utilities

The use of gas and electricity in an organisation is likely to be the second highest expenditure after staff costs. Increasingly, as sites become more sophisticated and acquire more technologies, electricity consumption increases as we move from naturally vented low-tech buildings to pressure-sealed environments that require mechanical cooling and ventilation. This is coupled with an increased reliance on computer technology, which adds to the cost of your utilities. New buildings, far from reducing your overall electricity consumption, may double it. You cannot really avoid this, unfortunately, but you can control the efficiency of all of the M & E equipment in the building. You also have some control over the proliferation of the other electrical devices that will inevitably find their way onto the site. Efficiency of M & E equipment is essential and you must try every possible way to reduce consumption through control measures implemented to optimise performance. Utilities and their procurement are discussed in more detail later in this chapter.

Waste

What waste streams does the organisation create? These could be as follows: general waste which cannot be recycled; recyclable waste, e.g. paper, card, packaging; plastics and tins; waste electrical equipment (WEE); specialist waste, the disposal of which is covered by specific pieces of legislation, such as batteries, fluorescent lighting tubes, oil and oil products and asbestos; and food waste. Waste is dealt with in detail in Chapter 9.

Transport

Note to self

When comparing data, make sure it is like-for-like and in a consistent form that you can easily understand.

Your organisation is likely to have its own transport, but you also need to look at how staff get to work and how schemes such as car sharing and cycling to work are promoted. In education, modern teaching methods mean that education is increasingly taken to the students, with off-site training and even international travel taking place to promote courses around the globe. All these areas need to be taken into consideration when calculating how transport affects your carbon footprint.

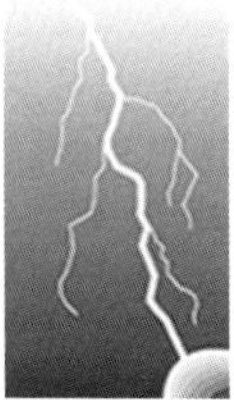

★★★★★★★★★★

Waste is often a large chunk of your expenditure in a busy organisation; don't forget to factor in your space requirements. Space costs money and all too often busy FM sections do not have the luxury of having a lot of space to store waste. Your solution needs to be minimalist while still being practical.

★★★★★★★★★★

Conversion of resource use into a carbon footprint

Converting your resources into a figure for the amount of carbon used is a tricky subject. There are countless methods to convert your resources used into an actual carbon figure (http://www.carbontrust.com/media/18223/ctl153_conversion_factors.pdf) Whichever method you use, stick with it, as changing the way you calculate your carbon use will invalidate your data. Using a consistent approach will help you to audit your figures more easily and assist your successors. Store your formulas on your PC and document how these formulas work and how they are constructed. There is nothing more frustrating than coming back to a spreadsheet after a year away and not understanding what you have done previously.

Setting bench-marks and targets

We need to know where we starting from when we begin our journey to reduce our carbon consumption. Discussing with your staff how you want to report your data is well worth the time taken and will help to provide consistency in reporting. Your data need to cover whole the site, but also need to be able to adapt as the site evolves, while still giving figures which can be compared over a number of years. Make sure you are comparing like with like. It is no use simply collecting, for example, data on energy used for heating each year without including data on outside temperatures to allow temperature differences to be factored out. Think in advance about what your data needs to represent and what factors could change that data and then it will be valid every year. Also decide how you want to visually represent your figures. Numerical data are fine, but pictures, graphs and pie charts are far more pleasing to the eye. Be careful about how you collect your data and make sure that you verify all your figures. If your calculations are flawed, then the final data will be too. You do not want to look stupid to senior managers, as this will immediately derail your attempts to create a low carbon environment. This will affect your credibility and make it less likely that you will get assistance in the future.

Setting future targets

You need to have targets to work to. Targets can sometimes be purely aspirations; this is an easy trap to fall into. If you set yourself difficult targets, you are setting yourself up for failure. You should always under-promise and over-perform. There is nothing more demoralising than to miss a target because you have been unrealistic. Set reasonable targets and exceed them if necessary. If you over-promise and under-perform, senior managers will be less inclined to believe any future targets you set. Use the SMARTER methodology for setting targets.

Reporting progress and reviewing results

You need to report your successes, particularly if they have created a financial saving. Use this to your best advantage, as some of your future projects may

Theory and information slot

SMARTER methodology:

S - Specific, Significant, Stretching

M - Measurable, Meaningful, Motivational

A - Agreed upon, Attainable, Achievable, Acceptable, Action-oriented

R - Realistic, Relevant, Reasonable, Rewarding, Results-oriented

T - Time-based, Timely, Tangible, Trackable

E - Evaluate

R - Re-evaluate

need capital funding and initial success means you are more likely to receive support from the finance director. Your reports should be simple and effective to achieve maximum effect. Avoid too much data. Dashboard-type reports seem to be the best and most eye-catching reports and these use a ragging (red, amber, green) system to report progress. You need to be honest in these reports as you can make your data say anything. Keep it simple and under-report as this then gives you more latitude the following year.

Sustainability: a whole organisation approach

Sustainability is a major topic for the practising Facilities Manager and can have a massive impact on your bottom line. If you ignore sustainability, you will suffer in the long term and put your organisation at risk. Engaging with the whole organisation is extremely important. You cannot have a successful sustainability strategy without engaging with staff and, in an education environment, students. Sustainability in an educational environment is different from that in any other type of organisation as sustainability is often proactively driven by the students rather than the organisation.

Staff need to understand some of the basics about sustainability before you can truly connect with them and influence their behaviour. The key areas include the following.

- *How much carbon do we create?* Staff need to understand the amount of carbon the organisation uses and how it is used. There is no point providing lots of data as this will only confuse people and your target audience will lose interest. Use simple bar or pie charts to show the relationship between consumption and carbon.
- *How much does carbon consumption cost?* Carbon consumption is a difficult concept to understand. We don't take the carbon away in trucks; it just gently floats off into the atmosphere as carbon dioxide emissions. If you represent your reductions in carbon emissions as potential financial savings, staff will easily see the connection between money and their actions, particularly in times of austerity when they can directly equate money wasted with potential job losses.

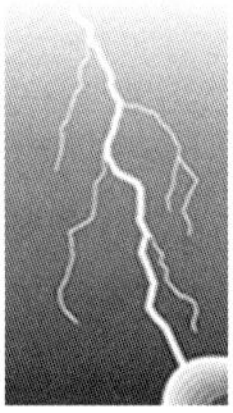

★★★★★★★★★★

The only way you will be successful with your sustainability strategy is if you get everyone on board. The best way to do this is to create incentive. Competition between departments with prizes works very well.

★★★★★★★★★★

- *What effect do carbon emissions have on staff?* How could the actions of staff reduce carbon emissions? It is no use asking staff to turn the heating down to save money. People all feel levels of cold differently and require a warm environment to work in; this is specified in employment law. Choose instead actions that you can influence, such a turning lights off, closing windows and turning off equipment when not in use. These are the simple things that staff feel they can contribute to and the areas they will, with gentle persuasion, comply with.

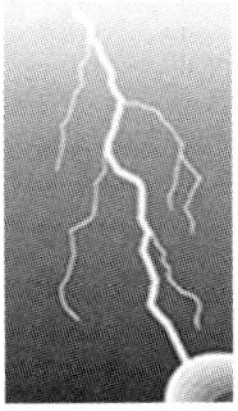

★★★★★★★★★★

One of the easiest ways to save money in your department is by adopting sustainable practices. Sustainability can pay for it self.

★★★★★★★★★★

- *Risks to the organisation.* You should inform your staff of the long-term risk to the organisation of inactivity. Sustainability will be one of the main drivers in the success of organisations in the future.

Note to self

The cost of utilities is going to spiral upwards over the next few years. Any savings you make from reduced electricity consumption will offset some uncomfortable decisions on budget reductions in the future.

Organisations which grasp the concept of sustainability fully will have a major competitive advantage and may find that success in this is rewarded by substantial gains in terms of funding, reductions in fuel consumption and kudos. It is likely that future funding in the education sector, in particular, will be biased towards institutions tackling their carbon emissions in a proactive way. Increasing fuel costs are almost inevitable. Organisations which proactively reduce their fuel consumption will reduce their ongoing running costs. Those who invest heavily in renewable technologies will see their reliance on the National Grid diminish and consequently will also reduce their vulnerability to rises in the price of electricity. Local communities often look to the organisations in their areas to be good examples of sustainability and this may give a competitive edge to organisations when attracting new customers and quality new staff.

An example of a sustainability policy is given in the Appendices.

Determining your carbon footprint

Note to self

Start with a simple list of all the items which you feel will contribute to your emissions. Remember you can't do this on your own; you need a Carbon Team to assist you. Once you have your list, divide it up into the various scopes. Try to keep to using scopes as this is the industry standard.

It is becoming increasingly important to determine a carbon footprint for your organisation; in fact, this is a requirement for some funding bodies. If used correctly, this information could give you a competitive edge. It can, however, create as many problems as it solves if not undertaken in a methodical fashion with defined outputs. Your carbon footprint is the total amount of carbon emitted by your organisation and, to be accurate, must be determined from current, auditable data. If you are determining your carbon footprint for the first time, you should take time to set the parameters for the data and decide how you are likely to keep and maintain it. It is more difficult to add data to an established system as inputting data to a system you have not created could cause problems with verification and accuracy, especially if your system for data collection is not clear.

The kinds of data you are likely to need for a typical establishment are as follows. You will need to define in advance which of these scopes you report; however, most organisations only report and monitor emissions from Scopes 1 and 2.

- *Scope 1 emissions.* Gas and fossil fuel emissions and F gas from air conditioning.
- *Scope 2 emissions.* Emissions from electricity used and bought-in heat.
- *Scope 3 emissions.* Emissions from transport, travel and commuting, waste disposal, and procurement. Water usage should also be considered.

Scope 1 emissions are your emissions from burning fossil fuels. Along with F gas emissions, these are relatively easy to calculate as your utilities suppliers should do this for you. Most, if not all, gas suppliers will provide these data directly to you, or they can be easily extracted from your fuel bills. If you have a procurement consultant, make them earn their commission by finding this information for you. The use of sustainable technologies to reduce carbon emissions is considered in detail later in this chapter. Scope 2 emissions are similarly easy to determine by asking your supplier or broker to do the work. Scope 3 emissions need a lot of work to be able to obtain a full and final figure. You need to look at the following areas.

Carbon emissions from transport

Business mileage can be broken down into business car mileage, train and bus fares, and air travel. You should record all your fleet mileage, vehicle by vehicle, so you calculate the carbon dioxide emissions by vehicle size. Make sure your system is overlaid onto a spreadsheet so that you can easily manipulate the data. For claimed business car mileage, make sure your claim forms record the distance travelled and the size of the car used. It will then be simpler and more accurate to plot your carbon dioxide emissions. For business train fares or buses, again record the distance travelled, not just the value of the rail fare. The same method should be used for air travel. For these data to be worthwhile, you need to be able to convert back to number of miles or kilometres travelled.

In an educational environment, it might be easier to find out the average distance travelled by students through a student survey. Make sure you break this down into journeys by public transport and those by car. Data about distances commuted by staff can also be obtained from staff surveys. If you think through how you want your answers returned, then these can do most of the work for you.

Carbon emissions from waste disposal

Calculating your carbon emissions from waste disposal will either be very easy or very onerous depending on how you currently dispose of your waste. To do this accurately you need to break your data down into each of the following waste streams.

- *Waste paper.* Is this separated, batched or baled, or does it simply go out as dry mixed recycling? Your paper will either be moved out by the skip load or by Euro Bin. Ask your collection service for data on the typical weights of both your skips/front-end loaders and Euro Bins, then simply multiply the number of lifts per year by your typical weight and you will have a good idea of the total weight of waste paper leaving the site. Alternatively, you may have a waste removal company which weighs your outgoing waste.

- *Waste cardboard.* If you have the storage space and the resource, waste cardboard is a saleable commodity and one that is worth batching and baling and taking off site for sale. Again, it should be relatively easy to calculate your total volume of card going off site.
- *Waste metal.* This is definitely a saleable item, but again requires a storage facility large enough to carry a metal skip. The market in metals is soaring at the moment, which you will no doubt be aware of if lead has gone missing from your roofs.

Water usage

It should be relatively simple to calculate your water usage from your utility bills; these will give you the number of cubic metres consumed. Saving water is far simpler than is often assumed. Sometimes water waste does not come from taps being left on, but from underground leaks that can go unnoticed. If your water is metered, and it is highly likely it is, then checking for leaks should be paramount in reducing your costs. The simplest way to do this is during a weekend close down when you are sure there is no one in the buildings. Make sure the systemisers on toilet areas are shut down and any water-using activities have stopped. Read the meters and then read them again after a couple of days before the building is in use again. If there is no difference between the two readings, then you can be relatively sure that you do not have a leak. However, if water usage registers, then you know that somewhere, probably buried out of sight, you may have a leaking pipe. Exploratory works can then usually find the leak. If you leave leaks, particularly underground leaks, then these are likely to undermine your building after a period of time. Therefore it is important to treat any discovery of a leak as a matter of urgency.

Another way to save water is in your toilet systems. Most old water systems have much larger toilet cisterns than actually needed. One easy way to reduce water consumption is to install a displacement device. There are many on the market, but most water boards recommend a product called a "buffalo." This is a simple liquid-filled sack which is dropped into the cistern, effectively reducing the amount of water required to flush.

Don't forget that your grey water should not be included in your carbon footprint exercise. From the perspective of sustainability this is free water, not water taken from reservoirs. Unfortunately, it cannot be used to offset carbon emissions either, unlike electricity generated on site. Remember that some aspects of grey water do expend energy; for example, pumping grey water around the building. One thing worth remembering is that your on-site water bill is split into two: one portion for the metered water coming on to the site and one notional cost for waste water going off site. It is assumed by the water authority that the same amount of water goes off site as comes on so your exported water is based off your imported (metered) consumption. If you are using your grey water rather than metered water to flush toilets then you are saving metered water and consequently your exported water (or water that goes down the drain) costs less.

Carbon emissions and procurement

This is always going to be a difficult area to assess. Unless you have all the data available to you in year one, then you may well find that improving your data as you go along will result in your overall carbon footprint going up rather than down.

Could you start to insist that your suppliers produce your goods with less packaging? There is no point in asking them to take packaging back with them as this hasn't stopped the packaging being produced in the first place. An audit of incoming goods will probably show just how much packaging is coming on to the site. You may find that most this packaging is needless anyway. A simple annual survey undertaken to determine what comes on to the site in one week will give you some useful figures. Other simple measures can be put in place to reduce packaging, such as asking your cleaners to reuse or refill trigger sprays rather than throwing them away. You will find that most chemical manufacturers provide chemicals in large bottles which can be decanted into your trigger sprays.

Do you know where your goods are being produced? Sometimes we buy things at the cheapest price without considering where they have been made. If you can buy the product more locally at the same price, the carbon footprint will be considerably less than a product purchased from the other side of the world.

How do you start to quantify the procurement practice accurately and does the amount of effort needed to achieve the final figures justify the process? This is a contentious issue and divides many people, but I would argue that, unless you have limitless resources, it is going to be difficult to come up with meaningful data that you can access easily each year.

What is it that we are trying to achieve overall with sustainable procurement? I would say generally that it is to reduce the amount of carbon we use and

> **Note to self**
>
> When choosing your team members, make sure you choose carefully and ensure that the person you choose has influence in the department or is able to make strategic decisions. Otherwise you are likely to stall the process when prompt decisions are required.

to reduce resource intensity. The best way of doing this is to raise awareness and try to manipulate performance. One good way of doing this is simply to request information from contractors and suppliers on their own performance. Most contractors and suppliers complete a PQQ when tendering for a contract. By adding questions about sustainability to the PQQ we can judge how well or badly they are doing in terms of sustainability. Simply monitoring and mentoring contractors and suppliers to become more environmentally friendly can improve your carbon footprint from the perspective of your procurement process. Setting up PQQs to give a scoring matrix can then indicate an improving sustainability rating and can be used on an annual basis to monitor performance.

Your "Carbon Team"

You cannot achieve this all on your own. You need assistance and your Carbon Team should be drawn from across your organisation from various disciplines. You should have made a list of where the carbon emissions are within the organisation and this should indicate the person most likely to be able to assist you. Your team needs specialists; for example, areas such as IT are difficult to tackle unless you have someone on board from the IT department.

If you are looking at paper usage, then it would be useful to have someone with a knowledge of reprographics involved in your discussions; similarly, areas such as procurement and catering should be involved. Your team must be prepared to put in the time and to take responsibility for some aspects of the process. You will find that lots of people volunteer their services, but that they are not prepared to undertake any work. They may cite issues such as pressure of work as their reasons for not continuing. Don't forget your students if you are in an education environment. They are the most likely to be able to drive processes forward and many students today know a lot about climate change and the effects of using finite resources.

> **Note to self**
>
> Pull together a simple PQQ to send out with all tenders. Make sure you are able to score your returns in a consistent way.

You need senior management to buy-in to the process early on. Buy-in by senior management will provide the necessary impetus to drive the process forward. Let's face it, some people are keen to get to the top. If they feel that senior management approve the process of carbon reduction, they are more likely to want to give up their time to gain management approval. There are also many members of staff who genuinely want to help with sustainability issues.

Organisational buy-in

Overall organisational buy-in is essential with respect to sustainability. Your carbon reduction plan needs to be structured so that you can report your progress and track your achievements. Organisational buy-in "rubber stamps" your process and makes it easier to get things done. Two key people are required in this process: a project sponsor and your finance director.

Warning
Don't set your targets too high. If you don't reach them, you will demoralise the team.

You will need someone on the executive at director level who can drive initiatives forward and ensure that you obtain the support you need to move projects forward. This project sponsor will also be your route to passing information through to the governors, who will be interested in any initiatives you may have. Buy-in from your finance director is imperative. Many of the projects you come up with will require a capital budget to get off the ground. Many projects will have a financial payback, but they will need that initial capital budget. If you can convince your finance director of your "spend to save" strategy you are more likely to obtain their buy-in for the whole process.

Recent problems with the economy have meant that many organisations are actively reducing their budgets and sustainability has a huge part to play in this. If you set yourself stringent but achievable targets for electricity consumption, using less electricity also means costing the organisation less. Many sustainability projects will give you substantial financial reductions as well.

Warning
Don't have too many in the team as this could bog you down.

Your carbon plan

You will inevitably need a route map to your potential success. You cannot expect to have measurable success on any of your projects if they are not mapped out and formed into some kind of plan to monitor your progress.

The essentials of your carbon plan are as follows.

- *A statement of the reasons why you are embarking on the road to sustainability.* These could include: to reduce the resource intensity of your organisation; to reduce running costs; to respond to peer pressure or pressure from students; to demonstrate your green credentials to customers; to comply with the Carbon Reduction Challenge; or simply to more effectively manage your carbon emissions.
- *Bench-mark values.* Basically, how have you performed to date? This will contain information about your emissions to date, such as the amount of electricity and gas used, how much water has been used, how much paper and card has been used, how much waste has left the site, how staff have travelled to work and what business mileage has been covered.
- *A summary of your targets.* For example, what percentage reduction have you targeted and how you are going to measure your progress?
- A summary of the projects you have identified and how these will influence your targets.
- The duration of the plan, e.g. when do you hope to reach your goal or planned percentage reduction?

Note to self

Remember to put the ethylene glycol replacement needed for ground source heat pumps into your planned programme. This can be an expensive item and, if you forget to budget for it, can be a nasty surprise.

Your plan needs to be published and time-bound and you need both staff and senior management buy-in. You need to report your progress to your management and governors and it is useful to have this as an agenda item at all management meetings to drive all the projects forward effectively.

Sustainable technologies

There are a number of technologies which call themselves "sustainable." Although this term is debatable, the following technologies do deserve consideration when you are trying to reduce your carbon usage.

To define sustainable technologies, you first need to define what the word sustainable means in the context of an organisation. I would define "sustainable" as any item which, in its preparation, execution or methodology, reduces the resource intensity of the way it is produced or manufactured, the way it is or can be delivered, or the way in which it is used.

A sustainable technology is any technology which adopts all of the above principles. Sustainability is a rapidly changing area, however, and what is sustainable today may be seen as archaic or resource-intense in the future. So when someone says a technology is sustainable, it could just be that it is marginally more sustainable than its predecessor. Sustainability therefore can be anything we choose and many manufacturers are already using and abusing the sustainability trend by branding products to suit fashion trends. For example, the term "eco" is often misused to represent items which are only marginally better for the environment than the previous version. It is usually up to the individual to make their own judgement about whether they are getting something which is better than the previous product.

Ground source heat pumps

Ground source heat pumps work by extracting solar heat stored in the soil or subsoil. The top 3 m of soil or subsoil maintain a near-constant temperature of between 10 and 16°C depending on latitude. At depths greater than 15 m, a constant temperature of 10°C is generally maintained. A ground source heat pump taps into this ready supply of heat in the ground by either laying a continuous horizontal trench, usually in a field or open area, or by directly sinking boreholes up to 100 m vertically into the ground. In my opinion, vertical boreholes are the better option as you don't need to worry about people digging them up. The pipes are generally filled with mono propylene glycol (MPG) to prevent freezing. These systems can also be used in reverse for cooling, but are generally limited to heat only. They are well suited to underfloor heating because of the lower temperatures needed.

Warning
Don't get talked into retrofitting a ground source heat pump as this is very costly unless you have a large field or lake to sink the boreholes in.

Care must be taken when specifying heating systems based on ground source heat pumps to keep the system as simple as possible. The simpler the system, the more effective and easier it is to maintain. As with any heating system, annual maintenance will be needed to keep the heat exchangers running correctly and to the right temperature; occasionally the system will need topping up with ethylene glycol. Every three to eight years, depending on use and the temperatures achieved, the ethylene glycol will need to be replaced to maintain efficiency.

The advantages of heating systems based on ground source heat pumps are as follows.

- Units can be dual purpose, heating in winter and cooling in summer.
- Geothermal heat pumps, although more expensive initially, can offer the greatest cost savings in the long term.
- They are mostly low maintenance.
- They offer the best solution for highly insulated buildings where thermal load is not large. They work well with underfloor heating.
- They can make a positive contribution to your Energy Performance Certificate and subsequent Display Energy Certificate.
- The estimated lifespan of the plant is 15 to 20 years.

The disadvantages of these systems are as follows.

- The initial installation cost is high, but can be more economical in a new build where boreholes can be incorporated into the building footprint. Retrofitting an existing building is usually uneconomic.
- Controls for the systems are not always user friendly.
- The heat pumps run for longer hours than traditional boiler systems.

Air source heat pumps

Warning
As most large organisations are urban or semi-urban, wind generation is not always viable.

Air source heat pumps are similar to ground source heat pumps in principle, but are generally marginally less efficient. Instead of drawing heat from relatively warm soil, air source heat pumps draw heat from the external air.

Biomass boilers

Biomass is considered to be carbon-neutral as it uses a renewable crop rather than burning a fossil fuel. It is not just about burning logs; many other fuels are used in biomass boilers, such as: energy crops, e.g. *Miscanthus*; agricultural waste; food waste; industrial waste; and woodchips and pellets. Most people consider biomass as an open-fire technology, although the truth is very different. With open fires most of the heat goes up the chimney and is lost. Biomass boilers convert as much as 70% of the fuel into useful heat. Pellet stoves can be even better, converting as much as 90% of the fuel into useful heat. Consideration needs to be given to how you store the fuels and their availability.

The advantages of biomass boilers are as follows.

- Biomass boilers now have much better controls than previously.
- Biomass fuels are considered to be a low carbon technology.
- The fuels used are sustainable.
- Fuels can be stored locally.
- Biomass boilers can reduce disposal costs as some waste can be burned in them.

The disadvantages are as follows.

- Fuel quality and cost can vary.
- The fuel has to be physically moved to the building.
- The ash produced has to be disposed of.
- Biomass boilers and fuel storage have a much bigger footprint than standard boilers.
- The market for this technology is still young and consequently sourcing of fuel can sometimes be difficult.
- You may fall foul of the Clean Air Act in some areas where the burning of smoke-emitting devices is prohibited.

Note to self

Take into consideration the size of boiler house and storage areas needed for biomass boilers. How will you obtain the fuel and is it a stable supply?

Solar thermal panels

Solar thermal panels are used to heat or preheat water to a temperature at which it can be used for domestic hand washing, or boosted for wider applications. Often confused with photovoltaic (PV) technology, solar thermal panels are completely different in that there is no generation of electricity. Solar thermal systems in the UK are generally "active," requiring pumps and cylinders. Indirect

Warning
Although very popular, the fuel supply for biomass is sometimes erratic and you may it find difficult to source suitable quantities of fuel.

or closed-loop systems use a heat exchanger that separates the potable water from the fluid, known as the heat transfer fluid, which circulates through the collectors. The two most common heat transfer fluids are water and an antifreeze/water mixture that typically uses non-toxic propylene glycol system. The better systems are designed to drain back when the pump stops working; this protects the system from frost and prevents excessive degradation of the propylene glycol.

Solar thermal panels require more space than PV technology because, in addition to a relatively large space on the roof for the collectors, space is needed in the plant room for the pump controller and heat exchanger. On a large project this can be a significant area. Don't forget to include the services of a structural engineer when you consider solar thermal as some of the equipment is heavy. You also need to consider the lifespan of your roof. The last thing you want to do is install a solar thermal system, then realise you need to replace the roof covering.

As with any heating system, solar thermal systems require annual maintenance to keep the heat exchangers running correctly and to the right temperature. Occasionally the system will need topping up with propylene glycol. The propylene glycol will need to be replaced every three to eight years depending on use and temperatures to maintain efficiency. Modern systems with drain-back fitted protect the propylene glycol, increasing its lifespan.

The advantages of solar thermal heating systems are as follows.

- Solar energy is available to all and free after the initial cost of equipment.
- A properly designed system can save you as much as 60 to 70% of your energy bills for hot water.
- Most systems can be made to pay back within ten years with feed-in tariffs.
- Sometimes grants are available to cover the cost of installation and interest free loan can be obtained. A good source is "Salix" funding.
- Solar energy has a positive effect on the environment and is safe to use.
- Relatively low maintenance is needed.
- Pumped systems minimise heat losses in the system.

Note to self

Try to look for sustainable technologies that have a payback of no more than six years. This way you can be sure of getting your money back over the period of most carbon reduction plans.

The disadvantages of solar thermal are as follows.

- The high initial cost, although there is a good return after investment payback periods.
- Performance depends on the weather; on cloudy days performance is likely to be low.
- It is not suitable for all properties and a south-facing aspect is required.
- An auxiliary heat source is needed when the system cannot perform due to weather conditions.
- The panels can adversely affect the appearance of the building.
- If the system is not pumped, there is a danger of freezing in winter.
- Active systems are more expensive to run than passive.

Photovoltaic panels

Photovoltaic technology has been around for a number of years, but has proved to be inefficient, resulting in long payback periods. This has always limited their purchase and installation to organisations with very strong environmental principles. However, today's modern panels tied to feed-in tariffs from the government have decreased payback periods to as low as nine years, although these feed-in tariffs are now being removed or reduced due to government cutbacks.

Photovoltaic panels can be fitted on the ground or on the roof of your property. They are installed at an angle of 30° from the horizontal and need to be south-facing for optimum performance. They require little additional equipment, generally just an inverter to convert from DC to AC.

Electricity produced by PV technology reduces the amount of energy used from non-renewable resources such as coal, gas, oil and nuclear power. In addition, there are significant environmental benefits from reductions in air pollution and carbon emissions from burning fossil fuels, reductions in water and land use by central generation plants and reductions in the storage of waste. Solar technologies produce energy with little noise and few moving parts.

There are currently three types of PV panel available and care must be taken to choose the right technology.

- *Monocrystalline.* These are manufactured from a single crystal of silicon and can be up to 20% efficient.

Note to self

Use the panel with the highest efficiency factor or, at least, do a calculation of cost in use.

- *Polycrystalline*. These are made from thin slices from a block of silicon crystals and can be up to 16% efficient.
- *Thin film*. These panels are made from a layer of semiconductor on a glass or metal base. They are the poorest performing panels at between 10 and 12% efficiency.

The advantages of using PV panels are as follows.

- They have no moving parts and no maintenance is required to the panels themselves; however, the inverters need to be maintained and will probably have a lifespan of between five and ten years.
- Electricity generated by solar power is produced without any greenhouse gas emissions and is therefore a high yield carbon reduction system.
- Little space is needed, especially in the plant room.
- PV panels are silent in operation.
- Solar energy will work in most environments as long as a clear south-facing aspect is available.
- There are many government initiatives available, although these are likely to decrease as the technology becomes more readily available. As a result, many schemes now have a reasonable payback period of between eight and ten years.
- Some companies will enter into agreements to supply the capital cost for installation providing they receive the feed-in tariff and you buy the electricity at a reduced rate.

Note to self

You need to check the structural stability of buildings before erecting a roof-mounted wind turbine to be sure that the building can absorb the additional lateral loads created.

The disadvantages include the following.

- The initial cost of installation and purchase is high compared with other technologies.
- Electricity production is dependent on the quality and quantity of light available.

- Electricity is only generated during daylight hours, so much less electricity is produced during the winter, which is the time of peak demand.
- The building needs a south-facing aspect and sufficient roof space.
- Long-term agreements with companies providing finance require commitments about access to the roof and roof life (some demand 25 year agreements).

Warning
PV panels are not economic and will not give you a payback on investment unless you get a feed-in tariff.

Wind generation

The areas which benefit most from wind generation are those on exposed sites where the wind is of sufficient strength and consistency to justify investment in a wind turbine. Small-scale turbines do work, but rarely merit the financial outlay needed in terms of electricity generated. In general, wind generation only works on a large scale with industrial sized turbines.

There are basically two types of turbine.

- *Turbines with a two- or three-bladed horizontal axis.* These need to face the wind to extract power and therefore need to be able to change direction with the wind.
- *Vertical turbines.* These do not need to change direction and are usually more suited to urban sites.

Voltage optimisation

The electricity which flows into your establishment does not always flow evenly. There are a number of factors that can affect the flow of electricity such as: harmonics, over-voltage, transients, phase imbalance and power dips.

Voltage optimisation is an electrical energy saving technique installed in series with the mains electricity supply to give an optimum supply voltage for the site's equipment. Voltage optimisation improves power quality on the system by balancing phase voltages and filtering harmonics and transients from the supply. Voltage optimisers are essentially electrical transformers: coils of metal that behave like magnets when an electric current is passed through them. Optimised supplies can therefore eliminate inconsistencies in the supply and remove the power surges that can cause so much aggravation and downtime in your IT systems.

Grey water

> **Warning**
> BMS systems need to be managed and run effectively. They do not run themselves and need technical back-up and servicing; this needs to be factored in.

Grey water is sometimes discussed along with other renewable technologies. Grey water captures a resource and re-uses it to the best effect. Grey water is water which has fallen onto a building and is captured through the roof drainage system and funnelled into holding tanks, generally underground. It is filtered, then pumped back into the building and held in storage tanks inside the building ready to be pumped around the building for use in flushing toilets. It does not produce potable water and therefore cannot be used for drinking water.

This system is sometimes known as rainwater harvesting and needs to be considered at the design and build stage. Retrofitting a grey water system is possible, but far more expensive. Maintenance is generally completely forgotten until the filters block and the system no longer works; it should be considered from day one.

Building Management Systems

A BMS is a standalone computer system that can calculate the pre-set requirements of the building and control any connected mechanical plant to meet the needs of the users of the building. The BMS does this by using inputs, such as temperature sensors and outputs, as on/off signals connected to controllers around the building. Programmes within these controllers use this information to determine the necessary level of control. The controllers are linked together and information can be passed from one to another. Connecting a modem allows centralised or remote access.

The level of control given via the BMS depends on the information received and the quality of information programmed into it. As well as offering a precise degree of control of its environment, it can trigger alarms about certain conditions and advise when a system is not operating to its maximum capacity. The systems are only as good as the people who are operating them and the information programmed into them. They need to be managed to work correctly and are not generally something that you can fit and forget. However, the benefits obtained from a well-run BMS make them a more than worthwhile investment. The kinds of systems that will run effectively from a BMS are: heating, ventilation, security door access and lighting. In general, if a piece of equipment has a mechanical signal, then the BMS can control it.

Installing sustainable technologies

> **Note to self**
>
> Organise all your staff to carry out a base load check one Friday afternoon. Log all meter readings, then close all systems down. On Monday morning, switch everything back on and re-read the meters. This will give you the base load.

When you first start to look at sustainable technologies, there is a tendency to look at each individual project in isolation. However, you should adopt a coherent strategy for sustainable technologies in terms of having a plan detailing your intentions throughout the site. You should have a structure to integrate these technologies into existing systems. Like all M & E systems, these new technologies need to be maintained and, in some circumstances, this maintenance can be onerous. This needs to be factored into your long-term maintenance strategy, otherwise you could find yourself saddled with considerable liability in the future.

You need to be able to measure the effectiveness of each of the items of technology that you install. There is no point installing PV panels if you don't know how much electricity they are generating. Equally, you need to be able to demonstrate that the claims you have made in relation to payback have been achieved. If you are able to do this effectively, then you are much more likely to receive continued funding for other projects. Make sure that these technologies are connected to your BMS so that you can readily draw down information to prove that the systems are working.

Utilities

One your biggest budget areas will be your electricity and gas consumption. Expenditure on electricity is often difficult to control without careful monitoring and re-education of your users. Gas consumption usually relates directly to heating and, as the amount of heating required depends on the season and outside temperatures, you have fewer options about how to make savings. However, before discussing methods to control consumption, you need to know how you currently use your electricity and what makes up your charges.

Meter readings can tell you how much gas or electricity you are using, but cannot tell you when you are using it. If you have any half-hourly meters, the data from these will be stored off site by your supplier and these data can be obtained on request. This will give you a load profile, i.e. a plotted graph indicating how much electricity is used at particular times of the day. This graph will indicate whether equipment has been left switched on when not required, e.g. in the middle of the night. Equipment commonly left switched on include pumps on timer switches, boilers, air conditioning, ventilation systems and heaters or water heaters with out-of-sync timer switches.

Establishing your base load

The first thing to establish is your base load. Your base load is the minimum amount of electricity needed to run your building when it is not in use. You cannot turn everything off in a building; computer servers and food freezers, for example, need to be on continuously. Pumps also may need to be left on and ventilation systems are often left on overnight for free cooling, so you will always be using some electricity.

The easiest way to determine your base load is to visit every building on a Friday afternoon and systematically turn off every appliance that has been left on unnecessarily. Make sure you discuss this with your IT department in case they have planned any downloads over the weekend. Turn off clocks, heaters, phone chargers and PCs, but make sure you leave any essential services on. Read the meters as late in the day as possible and then leave the building to work as normal. On Monday morning, read the meters again. This will tell you the amount of electricity used over a two-day period. Halve this value and you have your base load per day for your site. This seems rather simplistic, but it works. You can then establish a percentage reduction to aim for from your total consumption minus your base load.

Procurement of electricity

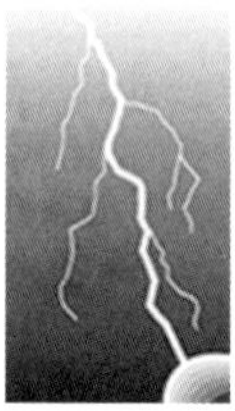

★★★★★★★★★

Procurement is a difficult area to deal with as it can clearly be very confusing. Use a professional consultant or engage with a consortium.

★★★★★★★★★

This exercise is usually carried out every two or three years. By the time the contract needs to be renewed, you have often forgotten everything you learnt last time you carried out the exercise. There are some essential pieces of information you need to gather before starting a procurement process.

- *Collate your sites.* Do you have addresses and details of each of the buildings on the contract?
- *Meter type.* Do you know the kind of meters that are in each building? It is useful to group them at this point into non-half-hourly and half-hourly meters.
- *MPANS.* Do you have all of your meter point reference numbers (MPANs)?

Procuring electricity has never been simple. Your electricity never comes directly from the generating plant to your doorstep. There are a number of players who all charge for their part of the process. These costs include the following.

- *Using the National Grid.* A sophisticated network brings electricity from the generating company across the country to your door. You sometimes see this described as Transmission Use of System (TUOS) and Distribution Use of System (DUOS).
- *Supplier costs.* Your supplier pays the power stations for generating the electricity and the National Grid for using their distribution system. This is sometimes seen on your bill as NGC or DNO.
- *Meter readings.* The meter operator charges for reading your meters.
- *Taxes.* These include the VAT element and the Climate Change Levy (CCL).

Buying and selling electricity is a risky business which depends on many factors; the market is extremely volatile at the best of times. The best organisations from the perspective of the sellers are those who are well organised, with a good understanding of the timing and amount of electricity consumption. Predictable usage and good data show that you know what you are talking about and you become a far more quantifiable risk. To achieve the best outcomes, ask yourself the following questions.

- What kind of meters do you have on site? If you have half-hourly meters, then you can provide valuable data to your supplier.
- Can you provide good consumption data in a format that your supplier can understand?
- Do you have all the necessary MPANs?
- Can you supply additional forecast data, e.g. are you likely to demolish a building or construct a new one?
- Have you made a choice about how long a contract you are prepared to take? One-year contracts generally give a better rate than a three-year contract, as there is more risk for the supplier in a longer contract.
- Have you set a target date for going to market and

Note to self

Pull together all your data from your meters and ask a meter specialist to give you a quote for AMR. It will pay for itself and will help you with both your CCL and sustainability targets.

are you flexible with this date? Remember, you do not always have to procure your next contract at the end of the current one. If the market is good, you can buy in advance at a more favourable rate.

- Are you able to aggregate your bills to ease administration? You can reduce your costs by putting them together in one bill.
- Do you understand what the best tariff for you is? Your may benefit from cheaper night-time tariffs?
- Can you compare suppliers on a like-for-like basis? If your requirements are complex, then this is when employing an energy specialist can pay off. If you do opt for an energy specialist, make sure that they are not tied into deals with particular suppliers.

Energy procurement is fraught with pitfalls. It can give you many sleepless nights, particularly if you are reaching the end of a good deal and the procurement process is going to result in a substantial increase in prices. This is a no-win situation, because your finance department is going to want to know why the utility costs have increased. The greater your understanding of the market and how it works, the easier it is to obtain the backing of your finance department.

Control of utilities

How do you manage your electricity and gas at the moment? The definition of control is the ability to monitor and influence. The first part of this control is to have an understanding of where your electricity is being used. In its simplest form, this is taking meter readings and using these readings to plot your usage on the site building by building. It is far better to use automated meter reading (AMR). In this system, pulsed sensors fitted to each meter record the electricity used and transfer the data a central point such as a website or your PC. Interpretive software uses the raw data from the meters to provide management information for your site. The better software provides models and interpretive reports which help with your energy analysis. If you have multiple sites, then investment in AMR is worthwhile if only because of the savings in staff costs from collecting meter readings. AMR is the cornerstone of good utilities management. Only when you can accurately measure your electricity consumption and carry out a comparative analysis can you truly control your use.

A good AMR system can help you with improved data collection and increases the performance of this. You avoid reading errors and missed meter readings. Reading data automatically also provides increased accuracy of data flow between the AMR system and other applications such as finance systems. Avoiding manual data entry or collection not only saves staff time, but will be far more accurate as you are not tied to collection on set days. With AMR you have constant access to real-time data and meter readings are available on request. Costs for meter reading will be reduced and therefore the total cost of the operation. Your data are stored so that you can manipulate them to provide comparative analysis.

The financial benefits of AMR last over the lifetime of the system. Your system will pay for itself over four to five years and any savings after that point are a bonus. You also save the time it takes for someone to read the meters manually. With AMR your utility bills are based on actual consumption. This generates a steady cash flow. Billing is based on real-time data and estimated bills are no longer necessary. You no longer need an interface between you and the utility company, i.e. no one needs to telephone the company with meter readings each month. You can also see how much gas and electricity individual buildings or departments are using. This helps with cost centres.

If you are aware of what equipment is using electricity and when, you can influence its use by turning things off that do not need to be on. You may find that, as faults have developed in equipment over the years, certain aspects of the system have been put “in hand,” i.e. taken off control and put into continuous mode. This happens in most FM departments as it removes initial problems prior to repair. Unfortunately, you sometimes find that these temporary fixes are forgotten and systems end up being left “in hand” permanently; this obviously uses more gas or electricity.

Timed systems can sometimes be set up incorrectly in the first place, with assumptions about usage being made and never challenged. A good example of this is ventilation systems set to come on a couple of hours before needed and then over running by a couple of hours at the end of the day. There could be a good reason for this, however, so do not turn things off without consultation.

Educating the occupants of buildings

The occupants of buildings need to be trained how to use their facilities. Without this basic training they can undermine your attempts to reduce consumption through their ignorance. A prime example of this is when staff bring portable heaters into work because they feel chilly in the morning rather than asking you to turn the heating on earlier. This can have a knock-on effect on electricity consumption if widespread. It can also affect the heating system if the portable heater is adjacent to a main heating sensor. This can

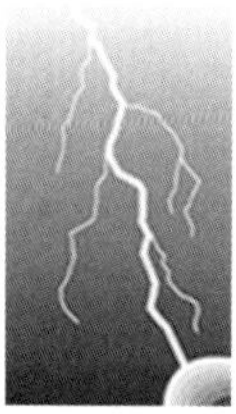

★★★★★★★★★★

A well-constructed and executed staff awareness campaign can save as much as 10% on your utilities costs.

★★★★★★★★★★

then turn the heating off in large areas because it has reached the required temperature in the local area. This create a vicious circle in that more people then use portable heaters and suddenly your building is being heated by portable heaters and you can't understand why your heating is not working correctly.

Staff throughout the organisation need to be on your side and must be aware of their actions in relation to energy use. An energy awareness campaign can help you to get the message over. Shock tactics work well, e.g. you can provide data on how much it costs to leave lights and PCs on in an office and how much it costs to boil a full kettle compared to boiling only enough for a single cup. This will pay dividends. You need to be able to check the effectiveness of the campaign, so before you start you should undertake a survey to check on staff perceptions about sustainability and energy usage. This will enable you to target areas of awareness that are clearly not up to scratch. It also saves you time and effort by not working on areas where staff knowledge is already high.

You must shout about your successes and publish your failures. This will have two effects. Firstly, it keeps the energy saving message alive as people can see that you are actually doing something and, secondly, it can shame people into action. Publishing your successes can provide support in itself. Success breeds success, and people want to be associated with success. You will find that the more successful your campaigns become, the more people will engage with what you are trying to achieve. Conversely, if you fail to get results, people will shy away from helping you as they will not want to be associated with failure. It is therefore important to get your staff awareness campaign right first time as you will be more likely to gain support and people will pay more attention to what you are doing.

If you are able to pull data together by building or, better still, by department, you can name and shame poorly performing areas. Setting up silly competitions like the "Green Wellie Award" may add a touch of humour to your campaign and engage people. No one likes to be seen as a "tree hugger," so the more you can steer away from this image the better. Making departments focus on their utilities costs will not only make them realise they are actually using electricity, but that they need to control their own usage and reduce consumption.

Opportunities to reduce energy costs will only arise if you have a consistent and long-term strategy on the way in which you engage with staff and students. Piecemeal campaigns will only work for short periods and your message needs to be forceful and continually visible. "Sustainability Weeks," where the focus is on the reduction of utilities consumption, help to focus attention on reduction, but you need to be forceful and you can only do this if senior managers are on board with your message. You may need to penalise poor performance, e.g. charging centres who leave PCs on in rooms when no one is present.

Aide Memoire: Sustainability and carbon management

- Try to identify the key elements of sustainability that will affect your institution, specially the financial elements.
- Make sure that when you buy new equipment or refurbish, you look at better quality, more efficient equipment.
- Find a consistent way of reporting your carbon data and stick with it.
- Look at your site with a fresh pair of eyes and try to define which sustainable technologies would work for you.
- Put together a carbon plan for your organisation.
- Look at AMR and AMT for your site to monitor and track your electricity and gas usage.
- Educate the building users on how best to reduce carbon consumption.

Chapter 9
Waste and how to deal with it

What is waste?

> **Note to self**
>
> Make a list of all possible waste streams on site, then compare this against measures you have in place to deal with the waste.

The waste generated by even a small site can be phenomenal. You will need to deal with this waste in the most cost-effective, the least labour-intensive, and the least space-intensive way. Waste can be defined as any item which is no longer of use to the site and needs to be discarded and removed from the site. Whatever strategy you decide to use for dealing with waste, your aims should be to

- reduce
- reuse
- recycle

You should track where all your waste has gone. A robust system of waste transfer notes, coupled with the use of quality waste removal companies, will ensure that you comply with legislation.

Waste can be divided into four main streams: paper, food, specialist waste and building waste.

Paper waste

This is probably the most significant waste stream on most sites in terms of volume, particularly in office and educational environments. There is usually a considerable traffic of paper going both into and out of the site. You need to identify the areas where this waste paper is generated. By showing your senior management the costs of both generating and disposing of waste paper, you may be able to reduce paper use in your organisation and therefore also reduce the costs of waste paper disposal and meet your sustainability targets.

Sources of paper

When we look at most organisations, we see vast amounts of paper being generated and, in turn, being turned into waste which needs to be removed from site. Although a lot of information is now captured digitally in office environments, the old-fashioned paper form seems set to stay with us forever. The validation given by a signature on a piece of paper probably means the use of paper for forms will never end.

Meetings are one of the great destroyers of our rainforests. How many meetings do you attend where vast amounts of paperwork are produced, despite the fact that we all know it will never actually be read? It will generally sit on someone's desk until there is no longer room for it and will then be binned. We need to question the culture, and indeed the necessity, of printing all these documents. Far better to send electronic links for the documents and to store them on shared areas where they can be accessed if required. Demonstrating the benefits of this way of working to your senior managers can be an uphill struggle, but if you can get top-down support for this you will make substantial progress in reducing your paper waste.

In an education environment, students' work produces much of your paper waste; sometimes it is unavoidable for notes to be printed. Perhaps we should all be living in a paper-free society because secure, safe technology is available for digital data management. However, many people have an inherent distrust of computers and still believe that it is easier to print documents. Some printing will inevitably always be necessary. However, advances such as Moodle and other student portals can help to capture and use information digitally without the need for printing, and these solutions should be encouraged. Many lecturers and teachers still dust off their notes each year and spend time before the start of term photocopying dog-eared, out-of-date notes instead of capturing them digitally. Many of these notes are immediately discarded by the lecturer or student, so you need to tackle this area as a priority. These are the people you need to convince to move to Moodle by explaining the benefits of the system. This will sometimes be an uphill struggle, particularly if your Information Learning Technology (ILT) strategy is not fully available across the site.

Cost of paper copies

Note to self

Try to link sustainability to your waste strategy. It can make a difference to sustainability targets. Recycling is one topic that is brought up time and time again by student groups.

This is difficult to quantify, but when considering how much paper you waste, you need to factor in a number of costs. The first cost to consider is the cost of the paper itself. This should be easy to find if you have a central point for procurement; it could be far more difficult if you need to consider individual sources of paper bought by resource managers in individual departments.

The cost of printing is split into a number of different areas, such as the cost of toner cartridges, the actual printers, electricity use and staff time. Toner cartridges can cost a phenomenal amount of money, perhaps £100,000 per year for even a medium-sized enterprise. If your finance department cost-centres toner cartridges, ask them how much is spent; you will be surprised.

If you can substantially reduce the amount of printing, it follows that you can also reduce the number of printers and the subsequent capital replacement of these. This can be tens of thousands of pounds. Printers consume electricity each time they are used. They are usually switched on early in the morning and remain switched on all day, whether you print something or not. We as a society have developed a culture of always wanting things ready to go. Printers will be using electricity in standby mode for at least the 37 hours per week that the office space is occupied, and more likely they are left on and consuming electricity 24 hours a day. You may as well just set fire to £10 notes! Even when powered down, but not turned off at the mains, they will be consuming electricity through vampire power. If networked, they will be consuming electricity through the network card. The only sure way to reduce electricity use is to turn equipment off at the electrical socket.

Are you able to quantify how much time is spent by staff simply getting the piece of paper to the printer? A likely chain of events might be as follows.

- *Procurement officer purchases the paper.* This doesn't happen automatically and the procurement officer's time has to be factored in.
- *Receipt of delivery to site.* Generally, this will go from the delivery point on site into a store, a task usually carried by site supervisors.

- *Reprographics.* If the work needed is printed rather than photocopied, then your reprographics staff will have a part to play.
- *Movement from store to local store in each department.* Again, this is probably carried out by your site supervision team.
- *Movement from the local store to printer.* This will probably be carried out by departmental technicians, or another staff member who will load the paper into the actual printer.

A similar process occurs with the replacement toner cartridges and, indeed, replacement printers. Finally, a good proportion of the finished article goes straight into the bin! This example illustrates that your paper chase is likely to cost a considerable amount in staff time alone.

Warning
Large amounts of paper or cardboard near buildings can be an open invitation to arsonists.

Waste paper

Much of your organisation's paperwork finds its way into the waste bin and becomes part of your waste stream. In addition, the volume of waste paper that can be produced during the normal course of working from other sources is considerable. This can include waste from the following sources.

- *Paperwork.* This may be brought in by staff and customers and then discarded.
- *Periodicals and papers.* Your library (in an education environment) and reception desks usually receive periodicals and papers every day and these will end up in the bin.
- *Leaflets.* Thousands of leaflets arrive on site during normal operations, from organisations offering support or information to leaflets from local takeaways offering your staff lunch options.
- *Junk mail.* We forget this most of the time, but consider the amount of junk mail you get at home. Multiply this by ten and then by the number of staff members you have and you will be getting somewhere near the total. Add to this the total periodicals from specialist areas and you end up with a lot of paper. You yourself will be swamped with at least a dozen different free periodicals each week, so it is easy to see the effect on your waste stream. Unlike with private residences, companies cannot opt out of receiving junk mail.

Twenty years ago there was a good market for waste paper and you could have sold it for a small profit. Unfortunately, the current market for recycled paper is non-existent and this is no longer a viable option. You should not send waste paper directly to landfill, so you need to consider how you store and ultimately dispose of this waste. The cheapest method is to simply load all your recyclable material together as dry mixed recycling. The advantage of dry mixed recycling is that you do not have several different recycling containers around the site.

You may, if you are lucky, find a local removal specialist who is prepared to take your paper waste off site without charge. This is the best option from a sustainability perspective as clean waste paper is easier to recycle into a new product than mixed waste, which needs to be separated. However, to make this a viable option you need to have the space to store your paper on site without it getting wet. Storing as bales is an option for cardboard products, but not for general paper as the bales often become too heavy to move. This can create problems with manual handling and it is far better to have an enclosed, lockable skip. Care must be taken when deciding where to site these containers as placing skips full of dry paper next to buildings is an open invitation to arsonists. Space may also be an issue. In today's strict world of space reduction, many organisations have trimmed both their internal and external space to reduce costs and maximise the use of their premises. This can present you with a logistical nightmare in terms of storage of waste, leading you to be far more innovative about how you get rid of your waste streams. When balancing your space issues, you may need an increased frequency of collection, rather than large banks of receptacles for waste.

Food waste

Food waste is becoming increasingly difficult to dispose of. It is becoming less favourable to put the waste directly into landfill and farmers no longer accept waste food for feeding pigs. There are many companies who will take your waste food away, sometimes at zero cost. One of the best ways to dispose of this waste is to send it to an anaerobic digester, which converts food waste into energy.

Anaerobic digestion is the natural breakdown of organic materials into methane, carbon dioxide and a compost residue. This takes place naturally, or can be accelerated and controlled in an anaerobic digester. A typical anaerobic digester is a sealed vessel, or series of vessels, in which bacteria act without oxygen. The organic material needs to be fully mixed and warmed. Mesophilic digestion takes place optimally around 30 to 38°C and thermophilic digestion takes place optimally around 49 to 57°C.

The mixture of gases formed during the anaerobic digestion of organic wastes is called biogas. It can be used in stationary engines to generate electricity, but is not suitable as a vehicle fuel. After removing the carbon dioxide, the remaining methane is known as renewable natural gas or biomethane. Biomethane is virtually identical to natural gas and has many uses. It is probably one of the most environmentally friendly sources of gas. The solid residues of the process are turned into compost, which can be incorporated into soil to improve fertility.

For many years, catering departments used macerators in their kitchens to cut down the volume of waste going to landfill. This practice is increasingly frowned upon by local environmental health departments as being environmentally unsound. Over a period of years, fat deposits will build up in pipe work and fat traps, which will clog the drainage pipes in the area. This is also becoming a food source for vermin and many urban areas are seeing an increase in the rat population due, in part, to this ready food source in the drainage system.

Specialist waste

Specialist waste includes batteries, toner cartridges, oils, fluorescent tubes and various sources of radiation.

Batteries and toner cartridges

All batteries and battery products should be separated from your general waste streams due to the toxic chemicals that they contain. Alkaline batteries can be safely disposed of with normal household waste. Never dispose of batteries in fire because they could explode.

It is important not to dispose of large amounts of alkaline batteries in a group. Used batteries are often not completely "dead". Grouping used batteries together can bring these "live" batteries into contact with one another, creating safety risks. Adapted from http://www.duracell.com/en-US/battery-care-disposal.jspx

Toner cartridges similarly contain toxic chemicals and many companies now operate collection services for empty toner cartridges and bottles. These are usually tied into rebates back to the user or may operate as donations to chosen charities. It is sometimes difficult to prove that the charities do actually receive the donations, but the principle is good. If your organisation is sufficiently large and you have a lot of printers, why not consider standardising

Note to self

Try to avoid using macerators in catering departments. Not only is this practice frowned upon by environmental health departments, but you will also need to set aside funds for cleaning fat traps and pipe work.

all your printers and employ someone, even in a part-time role, to refill the cartridges yourself? In a medium-sized organisation your annual toner bill could easily be more than £100,000, so this is a worthwhile objective and could give good returns. Standardisation of all your printers to a single type should easily be achievable across the organisation with a five-year time frame, given the speed of replacements in IT systems.

Oil waste

You will probably find that, particularly with a motor vehicle department in your portfolio, all oil waste products are separated and held in storage to be moved off site at regular intervals. It is also important that oil-contaminated products, such as oily rags, are collected separately and disposed off site in a controlled fashion. Motor vehicle parts may contain oil and cleaning fluids hold waste residues in solution. These also need to be carefully removed from site. Ensure that a full waste transfer audit system is in place.

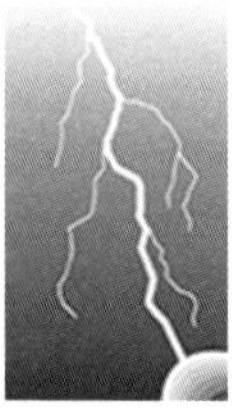

★★★★★★★★★★

When undertaking tenders or quotes for replacement lighting tubes and bulbs, make sure you specify the WEEE charge as a separate item. This way you can be sure it has been accounted for and you can check that you are not charged twice for the same service.

★★★★★★★★★★

Fluorescent tubes

For many organisations, disposing of fluorescent tubes and, increasingly, compact low energy light bulbs, is seen as an expensive problem that is often best delayed. One of the good things that came out of the Waste Electrical and Electronic Equipment Directive (WEEE Directive, 2002/96/EC) Regulations was an industry initiative to set up a free collection network. A levy of around 15 pence per tube is included in the purchase price of fluorescent tubes. This levy funds a scheme to provide containers and free collection to sites where there are likely to be more than 1000 lamps for disposal every 12 weeks. This scheme is operated through RECOLIGHT, a WEEE Compliance Scheme, who now have a network of disposal points across the country for businesses wanting to dispose of small numbers of tubes. Disposal through these points is free and, generally, you can just drop them off, although remember that when commercial waste lamps are moved from one site to another, there must be a consignment note for each movement. Be careful that you do not end up paying for your tube collections and remember you have already paid for disposal through your levy at purchase.

Radioactive sources

Any material which emits ionising radiation needs to be considered in relation to the various regulations. You may encounter the following radioactive sources.

- *Sealed sources.* A number of radionuclides, including radium-226, americium-241, strontium-90, cobalt-60 and plutonium-239, are used. Sealed sources require testing for leaks every two years.
- *Radioactive rocks.* Naturally occurring radioactive minerals, which have not been processed in any way, are often held by geology departments, sometimes without their radioactive properties being known.
- *Uranium and thorium compounds.* These are most often stored in sealed containers or puffer bottles and are used for half-life experiments.
- *Smoke detectors.* Radioactive versions contain a small americium-241 or radium-226 source.
- *X-ray equipment.* Any instrument generating X-rays will contain a radioactive source.
- *Neutron sources.* It is very rare for any organisation to own a neutron source. Registration with the Environment Agency is required, with high levels of supervision and documentation.

> **Note to self**
>
> More information on special wastes can be obtained from the Consortium of Local Education Authorities for the Provision of Science Services (CLEAPSS); most colleges and schools subscribe to this service.

Radiation sources need to be managed by a Radiation Protection Advisor. IRR99, The Ionising Radiations Regulations 1999 (S.I. 1999/3232), gives further guidance on this topic. All these sources need to be disposed of by specialist contractors and should never be put into standard waste streams.

Building waste

Building waste is probably one of the most bulky items that you need to remove from site and will depend on the size of the site and the type of building activities taking place. Three streams of building waste are likely.

- *Building waste.* Some departments, such as brickwork, joinery and plumbing, do try to reuse a lot of the items used in teaching; however, there will still be residual waste from items such as timber. You can only cut a piece of wood up so many times before you have to discard it and use a new piece. Similarly, brick and block work sometimes start to accumulate large quantities of waste. This should not be hidden away, but should be visible and integrated into the teaching process as an example of how to build sustainably and sympathetically reusing materials and disposing of waste appropriately. Suitable storage needs to be made available, usually adjacent to the construction area.
- *Minor works waste from the maintenance department.* Small jobs can create a vast amount of varied waste, from items such as old door handles and door closers to ceiling tiles and plasterboard, and, of course, empty paint tins. All this must be classed as building waste and, if possible, separated into its constituent parts for removal from site.
- *Major works or new build waste.* This will no doubt be written up in your Site Waste Management Plan. Modern building sites are usually set up to deal with waste in a cost-effective and environmentally friendly way. Buildings are designed to be built economically using standard sizes to reduce cut waste. Items containing gypsum, such as plasterboard offcuts, are collected and taken back to the gypsum factory for recycling. Metal offcuts are also usually separated to be removed off site.

Aide Memoire: Waste and how to deal with it

- Try to quantify your waste streams throughout your organisation.
- Recycle as much waste as possible.
- Make sure you fully document your waste streams in terms of removal and ensure you have the correct documentations for specialist waste or waste covered by specific legislation.
- Look at technology to replace some of the waste streams at source, such as Moodle and Papercut to reduce paper usage.

Chapter 10
Fleet management

Most FM departments will have a fleet responsibility and with this comes the obvious requirements to keep the fleet both maintained and legal. Your fleet may consist of just one van for the transport of maintenance materials around the site, or it could include a variety of vehicles from vans through to forklift trucks.

Whenever an area of your work is covered by legislation, it needs your full attention. It must be managed and documented fully to ensure that your liabilities are not only met, but are documented and can be audited. This links again to your QMS in terms of management and audit.

> **Note to self**
>
> Put together a full list of all of your drivers and set up a spreadsheet or database so you can track when your staff details need to be checked.

Your responsibilities fall into two main areas: the drivers and the vehicles.

Drivers and licences

You should not allow anyone to use your fleet without checking the authority and classification on their driving licence to ensure that they can legally drive the vehicle. For a standard van this could simply be the provision of a full driving licence. However, for a passenger-carrying minibus, a D1 classification on the driver's licence is the minimum requirement. It is, however, becoming far rarer to actually see this on a licence, as drivers who took their driving test after 1 January 1997 do not automatically get D1 on their licence. It is possible to take a D1 test to gain this classification and this involves both a theory and practical test and an enhanced medical examination. A D1 classification allows the driver to drive a minibus with fixed seating up to a maximum of

17 seats (including the driver), but only if this is not for hire or reward, i.e. it is on a voluntary basis. Hire or reward generally means any payment in cash or kind made by or on behalf of passengers which gives them the right to be carried in a vehicle. If you are paid for your role, and most FM staff are, it is far more difficult to use this form of licence legitimately.

To drive a minibus with nine or more passenger seats for hire or reward you need passenger carrying vehicle (PCV) entitlement. To obtain this, you must meet higher medical standards and take a further driving test. If you drive a minibus for a college or school under the minibus or community bus permit scheme, you do not need to have the higher PCV entitlement, although it is always good practice to do so. Minibus permits allow organisations to charge for travel in their minibuses without having to comply with the full public service vehicle operator licensing requirements and without the need for their drivers to have PCV (category D1 or D) entitlement. The service must be provided for their own members or for groups of people the organisation serves. The service must not be provided to the general public and the charges made must be on a non-profit basis. Community Bus Permits are issued to bodies who wish to run a local bus service on a voluntary non-profit basis using unpaid volunteer drivers. In this instance members of the general public can be carried in the minibus. It is difficult for an organisation to operate within the law using a permit.

It is far better, however, to insist that all staff who drive minibuses are fully PCV trained. This enables them to transport people legally and will be looked on far more favourably by your insurers. Details of the PCV course requirements can be found via your local authority or county council, who generally run these courses cheaply and economically.

Your first tasks are to establish the following.

- Who drives the minibuses.
- The licence status of the drivers, including taking a copy of the licence and keeping it on file.
- Are the drivers fit enough to drive? You must encourage staff to advise you of any medical condition that might affect their ability to drive.
- Advise your insurers of the drivers and their details.

You need to update all this information annually, so set up a computer-based system which requests it on an annual basis. Or, if your fleet is large, you may be able to invest in software to control this process for you. Remember the all-important insurance requirements. If you omit or make false claims about your drivers you will probably invalidate your insurance.

Set up some guidelines on how your drivers should use the minibuses. A series of user guides or Codes of Practice will be useful to your staff for specific scenarios, e.g. what to do if they break down, where they can obtain fuel, or guidance in the event of an accident.

Drivers' hours

The law on drivers' hours is complicated and, when dovetailed with European law, can become complex in interpretation. Three sources of legal information are:

- the European Union rules
- the Domestic GB rules
- the AETR rules (relating to a group of 17 countries outside the EU, mainly in southern and eastern Europe)

(European Union, EU Regulation 561/2006. Further information available at: http://www.dft.gov.uk/vosa/repository/Rules%20on%20Drivers%20Hours%20and%20Tachographs%20-%20Goods%20Vehicles%20in%20GB%20and%20Europe.pdf)

The definition of driving is being at the controls of a vehicle for the purposes of controlling its movement with the engine running. This applies whether it is moving or stationary. The total amount of time you are permitted to be on duty in the same 24-hour period is 11 hours. The Working Time Regulations 1998 will apply if you drive under the GB rules. These regulations set a maximum 48-hour working week, a right to 4.8 weeks of annual leave and a right to health checks and adequate rest.

When you are driving in the UK you must adhere to the following rules.

- Take a break of at least 30 minutes when you have been driving for 5.5 hours. Alternatively, within a period of 8.5 hours, breaks must add up to at least 45 minutes. This is so that you are not driving for more than 7hours 45 minutes. An additional break of 30 minutes at the end of this period must be taken to obtain refreshments, unless you are finished for the day.
- Ensure that in any working day the maximum amount of driving is 10 hours and that your total day's work does not exceed 16 hours.
- Take a continuous rest of 10 hours between two consecutive working days. You can reduce this to 8.5 hours up to three times a week.
- Have at least one period of 24 hours off-duty in any two consecutive weeks.

Vehicles

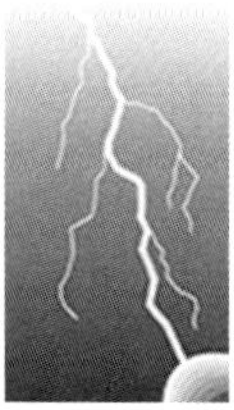

★★★★★★★★★★

Set up a spreadsheet to monitor the mileage of your fleet. This is good practice, not only from the perspective of tracking your fleet, but will also help you to define usage for your sustainability targets.

★★★★★★★★★★

Your vehicles need to be fully maintained to make sure you comply with your basic requirements for safety. Basic things such as MOT and service regimes must be fully managed. Simple spreadsheets can be compiled for these data. These can be linked to Outlook calendar reminders to ensure you are compliant with legislation and control your fleet adequately. Daily or weekly checks of the basics such as water and tyre pressures also need to be documented. Insurance cover should never be taken for granted and your insurers will be keen to see that you are covering all the basics in terms of management. This may well influence you insurance payments.

Key control

It is easy to lose control of your minibus fleet by not adequately controlling your keys. If you cannot control your keys using a formal booking in/booking out procedure, you may be open to abuse from staff taking vehicles for personal use. This may lead to invalidation of your insurance policy. There are many proprietary systems that can control your keys automatically through software and strong box solutions.

Logging journeys

Note to self

Set up a spreadsheet for all your vehicles and enter details such as MOT due dates, insurance renewals and service dates. Also put the key dates into your Outlook diary to make sure you don't miss them. You will not look at the spreadsheet every day, but you will look at you diary daily.

It is sensible and practical to log the mileage undertaken by your staff in vehicles owned by your institution for the following reasons.

- To control the use of the vehicle. When the mileage is logged, you have control. Without control, you cannot say where the vehicle has gone or whether it was on official business. This can only lead to misuse.

- To help you decide service frequencies and maintenance regimes.
- When you have a number of vehicles, it allows you to rotate them to obtain consistent use, instead of wearing out one particular vehicle.
- To help you to calculate your carbon footprint or make assessments of budget provision for future years.
- To help you decide which vehicles need to be replaced under your capital strategy.

Trackers

Use of a tracker clearly identifies when and where a vehicle is used and is a strong deterrent to misuse. Be sure that your tracker is installed in an inaccessible place in the vehicle, otherwise it may be disconnected. Larger vehicles may have tachographs which monitor vehicle use, shift times and, importantly, the break times of the drivers. Driver compliance is paramount and legislation around transportation must be followed precisely.

Fuel cards

This depends on the size of your fleet; however, in my opinion, fuel cards are essential if you want to control your vehicle usage. Most fuel cards have moved from allowing you to fill up at the local petrol station on a daily basis to the flexibility of using a fuel card with a major national fuel provider.

Capital purchases

Use the same principles as for your planned maintenance programme and apply them to fleet management. In this way you can predict in good time when you need to replace your vehicles and you can give an overview to senior management of usage against cost.

Aide Memoire: Fleet management

- Ensure all your insurance details are up to date.
- Ensure all your drivers are eligible to drive the vehicles requested.
- Set up spreadsheets to monitor items such as MOTs and service regimes.
- Set up a fuel card.
- Install trackers on your vehicles if you think they are likely to get misused.
- Check the European and Domestic rules on driving.

Appendix I

Strategic objectives and operational targets for a typical college

Appendix IA Helpdesk

Strategic objective	Action	Target 1	Target 2
1. Provide a vibrant learning environment to meet the needs and expectations of the College community	Be first contact to all users of the building. Provide frontline security to the premises and check all visitors, staff and students have authority to enter the premises	Undertake a monthly customer care report personally asking staff, students or visitors for opinions	
	Provide rooming reports through formal walk rounds and checks	Enter data from the reports onto the Helpdesk	Follow up rooming reports in terms of progress
	Make sure Helpdesk functions correctly	Undertake a monthly appraisal of current system and report back to Helpdesk software company on rectifications needed	Arrange an annual meeting with Helpdesk software company discuss problems
	Sort all outgoing mail and frank post daily	Undertake a monthly report on how much post has been sent out and how much it has cost	Break down postage out by curriculum/cost centres
	Receive, check and distribute incoming goods	Make sure all deliveries leave the post room at the end of each day on the reprographics/parcel run	Provide a quarterly report on number of parcels out to centres; aim for 95% completion rate

2. Ensure accountability and financial stability	Provide telephone assistance to clients by receiving Helpdesk work requests. Input all required data to progress the work requested. Provide day-to-day administration in support of the Helpdesk. Process urgent requests via radio and email. Process non-Helpdesk tasks such as porterage, waste transfer, room setup, cleaning requests. Report problems to contractors such as lift and fire alarm problems on Helpdesk issues. Attend disabled toilet and lift alarms where required. Liaise with the on-site security service	Provide monthly and quarterly reports on Helpdesk output	Monitor target completions for tasks in the Helpdesk
	Track number of jobs placed with each discipline within the FM department	Provide monthly list of number of jobs by trade	Move towards adding in time allocation to each job to link to trades with the ability to show a monthly total of hours worked by each team member
	Undertake return to work interviews for each member of staff in team	Undertake and keep a formal record of all staff sickness	Report on a termly basis to Facilities Manager

(*Continued*)

Appendix IA (*Continued*)

Strategic objective	Action	Target 1	Target 2
3. Set high expectations to promote the consistent achievement of aspirational learning goals	Secure keys in cabinet. Report damage and vandalism both internally and externally	Maintain a key book for contractors ensuring all contractors who come on site are logged in	
	Maintain evel of CPD within department	Hold and maintain a full record of CPD undertaken. using as a prompt for organising staff training	
4. Take action to ensure a safe environment. Promote individual understanding and responsibility for safety and well-being	During rooming report run check for access and egress issues	Place all works onto Helpdesk as and when rooming reports have been compiled	Recheck following month to ensure works have been undertaken
5. Promote equality actively, support diversity and cohesion and tackle discrimination through innovative strategies	Provide staff training in customer care	Arrange and organise an annual update for all staff	Assess effectiveness through appraisal process and amend as needed to ensure effectiveness

	Provide staff training on new Helpdesk software	Arrange and organise an annual update for all staff	Assess effectiveness through appraisal process and amend as needed to ensure effectiveness
	Provide staff training on manual handling	Arrange and organise an annual update for all staff	Assess effectiveness through appraisal process and amend as needed to ensure effectiveness
	Provide staff training on the franking of post	Arrange and organise an annual update for all key staff	Assess effectiveness through appraisal process and amend as needed to ensure effectiveness
	Ensure a full meeting schedule has been completed	Enter all meeting dates on Outlook calendar	Publish all dates to ensure people are aware of meetings
6. Develop a clear and open communication and engagement strategy	Encourage two-way communication within the Helpdesk section	Undertake monthly meetings which must be minuted; all minutes to be sent to Facilities Manager	
	Allow curriculum centres to engage with the Helpdesk by inviting resource managers into the Helpdesk to see how best we can serve them	Arrange regular one-to-one meetings with all members of staff	

(Continued)

Appendix IA (*Continued*)

Strategic objective	Action	Target 1	Target 2
7. Provide a range of learning opportunities and services relevant to the needs of individuals and employers	Encourage cross department liaisons and working arrangements to allow crossover of roles	Have an open day when key members of curriculum centre staff can drop in and obtain an understanding of how the Helpdesk and Postal systems work	
8. Provide a supportive environment for all staff encouraging personal well-being and professional development	Encourage FM team to work together	Arrange visits and close liaison with other FM departments to allow upskilling and succession possibilities	Encourage cross service working e.g. buddying up for one day per month with another section
	Encourage team events	Pull together a set of team events for the year, e.g. team barbeque, liaise with other teams to ensure full team involvement	
	Encourage paperless systems of working within the department	Encourage the team to look at ways of digitising items such as surveys	Work with the team to collectively work on 'paperlite' solutions for the department

9. Deliver sustainable education; reduce the resource intensity of learning with the long-term objective to have a carbon neutral site	Encourage others not to print documents or emails	Aim to reduce paper use within Helpdesk area by 10% year on year
	Reduce the amount of electricity used in the Helpdesk department	Try to persuade staff not to print emails
	Help coordinate the college's sustainability message	Encourage the whole Helpdesk team to completely switch off PCs and other post room equipment at the end of each day
	Ask your team to push the sustainability message at every opportunity	Include a sustainability message on customer satisfaction surveys

Appendix IB Management

Strategic objective	Action	Target 1	Target 2
2. Ensure accountability and financial stability	Maintain financial stability for the FM department	Estates to maintain +/-3% leverage on budgets held	Budgets which overspend to be micro-managed to reduce to the +/-3%
	Ensure all contractors using the college are sound and that full records are maintained for each contractor	All contractors to be written to annually to update details	Check all records at month 6 to ensure compliance
	All contracts to be re-tendered every three years	Maintain a list of current contracts and prepare and re-tender each contract on schedule	Review specification at each re-tender
	Undertake return to work interviews for each member of staff in team	Undertake and keep a formal record of all staff sickness	Work with HR to reduce sickness levels
4. Take action to ensure a safe environment. Promote individual understanding and responsibility for safety and well-being	Ensure the site is supplied with full CCTV coverage	Maintain the CCTV team and ensure at least a termly meeting of all parties including the supplier discuss technical difficulties, minuting all meetings	Increase the overall coverage of CCTV on site by detailing a plan giving site coverage

	Develop and maintain a lone worker system on site	Identify key lone worker positions and work with health and safety manager to complete a strategy	Ensure strategy has been followed by undertaking sample checks at month 6
	Ensure a strategy is in place for heating and cooling for the entire site, including a BMS strategy	Complete and update the BMS and heating strategy for the site and publish findings	Check with curriculum centres at key points of year to ensure they are happy with temperatures
	Ensure a PPM strategy is in place and working	Continually update the PPM structure to accommodate new liabilities	Check effectiveness of the completed schedule and check completed rates
5. Promote equality actively, support diversity and cohesion and tackle discrimination through innovative strategies	Ensure all contractors are vetted to ensure CRB checks are undertaken	All contractors who come onto site to be CRB checked	Annual update required from contractors to ensure all staff who come on site are fully CRB checked
6. Develop a clear and open communication and engagement strategy	Ensure a meeting structure is in place to provide effective dissemination of information to all staff	Place all meetings on Outlook calendar	Provide full set of minutes and agendas for full year
	Allow curriculum to use all FM facilities and resources to provide curriculum, e.g. use of boiler house for teaching	Ensure a termly meeting is arranged with all curriculum centre managers	Ensure an annual meeting with all resource managers

(*Continued*)

Appendix IB (*Continued*)

Strategic objective	Action	Target 1	Target 2
7. Provide a range of learning opportunities and services relevant to the needs of individuals and employers	Encourage a wider use of facilities for teaching	Work closely with curriculum centres to develop the use of the site renewable technologies for teaching	Ensure risk assessments complete for all facilities
8. Provide a supportive environment for all staff encouraging personal well-being and professional development	Provide a development structure for all staff to ensure ongoing and updated training	Annually review outside the appraisal process	
	Provide a methodology to ensure succession within the department	Compile a skills register for the entire team and encourage staff to take on crossover skills	Through appraisal process, try to identify areas not currently undertaken and challenge staff to train for these
	Ensure regular one-to-one meetings with all staff at least twice per year outside appraisal process	Encourage all staff to step up when opportunity arises by developing training (unpaid) for any team member who wants to learn about other areas	
	Try to enable staff access to managers	Encourage all managers/ supervisors to keep an open door policy for all staff	

9. Deliver sustainable education; reduce the resource intensity of learning, with the long-term objective to have a carbon neutral Site	Engage with students and staff to encourage reduced utilities on site	Attend student meetings to obtain feedback on sustainability issues	Encourage student sustainability groups to become involved with the college's Carbon Plan
	Ensure all contracts have been through the sustainable procurement process	Try to develop a communications strategy for the FM department to take the lead on sustainability	Ensure the PQQ for sustainability is used for all contracts
	Maintain a league table of performance for each building	Use information from AMR o create a league table	Publish the league table each term

Appendix IC Maintenance

Strategic objective	Action	Target 1	Target 2
1. Provide a vibrant learning environment to meet the needs and expectations of the college community	Maintain the grounds to provide a safe and vibrant environment	All bins to be emptied once each week and high use bins emptied more frequently	All smoking shelters to be cleared on a daily basis
	Ensure all areas are painted and paint finish is maintained	A monthly dilapidation report detailing where touch ups are needed on site	Ensure that larger works are scheduled taking account of curriculum needs
	Ensure all internal plant displays are maintained and kept in tip-top condition	Ensure that all plants are fed at least weekly throughout the summer	If needed for special functions buy-in/rent plant displays
2. Ensure accountability and financial stability	Internal ordering processes to be in place for minor materials	All orders to be formally generated	All orders to show due diligence and follow financial regulations
	Encourage internal buying of services, e.g. fully priced works such as shelves and minor improvements	All internal orders to be formally costed and paid for by curriculum centres	Ensure that internal invoicing takes place
	Undertake return to work interviews for each member of staff in team	Undertake and keep a formal record of all staff sickness	Report on a termly basis to Facilities Manager

3. Set high expectations to promote the consistent achievement of aspirational learning goals	Ensure the skills base for the department is detailed and training is up to date	Create a skills competency record for the whole team including site supervisors and be responsible for actioning and coordinating training in relation to maintenance tasks only	All team to be fully trained by end of year 1
	Detail how the maintenance department works and publicise this throughout the FM department	Provide a PowerPoint presentation on the maintenance setup: what we do, how we do it and what we stand for	Create a training programme for curriculum centres during staff development week
4. Take action to ensure a safe environment. Promote individual understanding and responsibility for safety and well-being	Ensure all access and egress is maintained at all times and alternatives are maintained and published	Ensure the groundsman is detailed with keeping all access and egress areas clear and that this is monitored and documented, e.g. a tick list which is date-stamped	Ensure snow plan is maintained and stocks are ready for first snowfall and that stock levels are maintained throughout the winter
	Maintain an acceptable predefined level of lighting on site at night	Undertake a termly check on all external lighting, carried out and detailed by contractor	Ensure any faults found are actioned on the Helpdesk

(*Continued*)

Appendix IC (*Continued*)

Strategic objective	Action	Target 1	Target 2
5. Promote equality actively, support diversity and cohesion and tackle discrimination through innovative strategies	Allow team to undertake enhanced training on safeguarding to counter stereotyping	Put together enhanced training on safeguarding to dispel poor image of construction staff	Encourage positive image to escape the constructor stereotype
	Organise awareness sessions on equality and diversity	Include at least one annual session on equality and diversity documented on training records	
6. Develop a clear and open communication and engagement strategy	Complete a full team meeting structure for the full year	All meetings to be minuted and sent to Facilities Manager	Encourage one-to-one meetings at regular intervals throughout the year fcr all maintenance staff
	Encourage a two-way dialogue between curriculum centres and maintenance to pick up perceived problems before they become an issue	Encourage a drop in once per term with each curriculum centre manager	Undertake annual customer satisfaction surveys on comfort levels in buildings
7. Provide a range of learning opportunities and services relevant to the needs of individuals and employers	Provide staff training in specialist maintenance trades in the form of toolbox talks	Enhance skills throughout the department by building on the competency listing to create an enhanced or supplementary skills listing	

	Ensure training on the BMS system for key personnel	Arrange full training on the BMS system to any new plumber	Set targets for new plumber on BMS, e.g. reduced failure rate of pumps through active management
8. Provide a supportive environment for all staff encouraging personal well-being and professional development	Encourage team events	Pull together a set of team events for the year, e.g. team barbeque, liaise with other teams to ensure full involvement	
	Ensure all staff attend occupational health monitoring set up for them	Ensure attendance	Ensure any staff concerns are raised with occupational health team
	Ensure that return to work interviews are carried out for every member of staff	Aim for 100% documented return to work interviews with a copy held on staff file	
9. Deliver sustainable education; reduce the resource intensity of learning, with the long-term objective to have a carbon neutral site	Ensure items are recycled and reused	Maintain a register of all reused items for the sustainability agenda and assess savings made in terms of both money and carbon saved	Aim for 50% reuse of old materials
	Construct and manufacture using standard sized items to save on offcuts	Buy only appropriately sized sheets rather than cutting down	Design items with standard sheet sizes in mind
	Purchase sustainable maintenance materials	Ensure that materials are sourced locally to improve sustainability	Encourage local suppliers to provide the right materials

Appendix ID Site supervisors

Strategic objective	Action	Department	Target 1	Target 2
1. Provide a vibrant learning environment to meet the needs and expectations of the college community	Provide assistance to curriculum centres in terms of setting up and removing displays and classroom layouts		Provide a termly report on how well setups have performed over the term and seek to find better ways to provide service	Obtain feedback from curriculum centres
	Ensure internal windows and foyer entrances are cleaned on a regular basis		Clean all entrance glassware once each week	Ensure this is entered on Helpdesk as job or PPM
	Set up examination rooms in line with requests		Ensure close liaison with relevant departments on setups and report back termly on number of successful exam setups; aim for 100% completion rate	
2. Ensure accountability and financial stability	Monitor and manage all minor stock items such as lighting tubes		Produce a termly updated stock list	Ensure procurement is undertaken to replenish stock
	Ensure timesheets are filled in timely and accurately		Ensure 100% completion of timesheets within first two days of each month	Aim for 0% errors on timesheets

	Ensure all Helpdesk jobs are actioned on a daily basis and that they are closed down accurately and timely	Aim to have at least 90% successful completion of all jobs	Quality control remaining 10% to find out why they were not completed
	Undertake reprographics run on a daily basis	Ensure 100% completion of run each day	
	Open and close each building in accordance with published opening and closing times	Ensure a 100% completion on buildings opened within the first 20 minutes of opening	Establish back up procedures to ensure buildings are opened on time
	Ensure coverage in terms of availability of staff	Ensure acceptable level of staff at all times by monitoring both sickness and holidays and scheduling accordingly	Schedule staff from other sections to ensure minimum staff attendance achieved
	Undertake return to work interviews for each member of staff in team	Keep a formal record of all staff sickness	Report on a termly basis to Facilities Manager
3. Set high expectations to promote the consistent achievement of aspirational learning goals	Ensure porterage of materials around site is quick and efficient to ensure that curriculum staff can operate seamlessly	Ensure a 100% target on all porterage	Sample quality checks to ensure compliance
	Set and monitor internal staff training and ensure all are undertaken each year	Liaise with FM office on training schedules	Ensure 100% on mandatory training across FM department

(*Continued*)

Appendix ID (*Continued*)

Strategic objective	Action	Department	Target 1	Target 2
4. Take action to ensure a safe environment. Promote individual understanding and responsibility for safety and well-being	Carry out rooming reports on a monthly basis, checking all tables, chairs etc.		Ensure all rooming reports are passed through to the Helpdesk for action	Ensure any defective equipment/furniture is removed from service as soon as possible
	Undertake regular checks on access and egress; monitoring reports to be held in central file		Maintain report on access and egress with formal checklists identifying when areas surveyed and by whom	Ensure issues are dealt with promptly
	Undertake a weekly check on external lighting when dusk is before 9 p.m.		Maintain a report on lighting around site detailing when checked and by whom	Undertake annual report on lighting, taking into account feedback from curriculum centres after open evening.
	Remove waste materials on a daily basis to the nearest waste disposal point		Ensure waste materials are where could cause or contribute to an accident	
	Provide operational briefs prior to holiday periods to allow colleagues to undertake works effectively		Ensure that a full list of current jobs and actions is passed on to next in line to ensure completion	Provide a hand-back brief when person resumes duties

5. Promote equality actively, support diversity and cohesion and tackle discrimination through innovative strategies	Encourage staff to report issues through line managers	Operate a 0% tolerance of equality and diversity issues	
6. Develop a clear and open communication and engagement strategy	Ensure a monthly meeting is undertaken and that the meeting is minuted	Book in a formal monthly minuted meeting	Circulate all minutes
	Ensure staff can talk on a one-to-one basis with their line manager	Ensure that a termly individual chat is undertaken with all staff and any issues raised brought to team meetings	Ensure issues raised are linked into appraisal
	Ensure all feedback on Helpdesk jobs is thorough and conclusive	Ensure recording of Helpdesk jobs on the software package	
7. Provide a range of learning opportunities and services relevant to the needs of individuals and employers	Ensure current skills in relation to Evac chair is up to date and that all site supervisors are fully conversant with their requirements in the event of a fire or bomb threat	Ensure this goes on the competency record	Ensure this training is also linked into a Personal Emergency Evacuation Plan (PEEPs)

(*Continued*)

Appendix ID (*Continued*)

Strategic objective	Action	Department	Target 1	Target 2
8. Provide a supportive environment for all staff encouraging personal well-being and professional development	Monitor staff to identify skills shortfalls and encourage staff to undertake relevant training to upgrade skills		Ensure all staff fill in the competency sheets identifying skills levels and work with the Helpdesk team to ensure all staff are upskilled	Ensure 100% completion of competency skills and that no staff member operates in areas that they have not been trained in
	Encourage team events		Pull together a set of team events for the year, e.g. team barbeque, liaise with other teams to ensure full involvement	Request ideas for an inclusive team event
9. Deliver sustainable education; reduce the resource intensity of learning, with the long-term objective to have a carbon neutral site	Ensure all waste is taken to the appropriate waste disposal point and is placed in appropriate waste streams, attempting to recycle as much as is practical		Collect daily all waste from around the site and deliver to appropriate waste disposal points, separating waste when possible	

Turn off lights in rooms on night shift as early as possible	Turn all lights off as the evening shutdown progresses
Turn off computers in unused rooms during close down period at night	Adopt a 'turn it off if there is no one in the room' mentality
Ensure the college van is filled with biodiesel	Aim to use at least 80% biodiesel as the year progresses

Appendix 1E Environmental issues

Strategic objective	Action	Department	Target 1	Target 2
1. Provide a vibrant learning environment to meet the needs and expectations of the college community	Provide waste information to college community		Provide termly records of waste moved off site	Feed in targets to the carbon plan
	Ensure the waste strategy is capable of moving waste away from learning areas quickly to promote a clean environment		Review as necessary to ensure compliance	
	Ensure that recycle stations are set up in each major office area to allow active recycling		Aim to have full recycling stations set up by end of year when new building comes on stream	Promote positive recycling with the curriculum centres
2. Ensure accountability and financial stability	Control college's EMS ensuring ISO 9001 and 14001 compliance		Ongoing	
	Control stock levels in domestic stores		Ensure that stock levels are maintained by periodic checks against stock. Feed back stock level replenishment to administration	Ensure stock is procured in a sustainable way
	Log out all WEEE and keep details of all equipment scrapped		Keep records of all disposal of electrical equipment	Look to positively repair WEEE rather than scrap
	Undertake return to work interviews for each member of staff in team		Undertake and keep a formal record of all staff sickness	Report on a termly basis to Facilities Manager

3. Set high expectations to promote the consistent achievement of aspirational learning goals	Ensure ongoing training for all environmental staff to keep skill levels up to date	Update annually	
	Identify any further CPD required for the environmental section	Update annually	
4. Take action to ensure a safe environment. Promote individual understanding and responsibility for safety and well-being	Ensure PAT testing is undertaken	Provide a monthly update on progress and identify slides in programme	Ensure any slides in programme are filled by external contractors
	Undertake fire extinguisher checks	Undertake on a termly basis for each building	
	Undertake fire alarm tests	Undertake on a weekly basis in all areas	Feed results back into annual statistics on compliance
	Provide access and egress reports	Provide termly report on progress	Feed results back into annual statistics on compliance
	Ensure all failed PAT test items are removed from circulation and are either repaired or removed to WEEE store	Remove all redundant or failed items to WEE store	Feed results back into annual statistics on compliance

(Continued)

Appendix IE (*Continued*)

Strategic objective	Action	Department	Target 1	Target 2
	Identify in advance trends for equipment failure and report these to curriculum managers		Report back any trends in failure to centre manager, e.g. if a number of hairdryers have failed, this could indicate the whole batch is about to fail and would allow planning of budget for replacement	
	Encourage staff to use the FM services for disposal of all redundant waste and COSHH related items		Periodically inform all staff of the need to dispose of chemicals safely through the FM department	
	Encourage participation in the weekly fire drills and termly evacuations by working with the health and Safety Manager to provide guidance notes		Monitor and formally log all fire alarm tests and fire evacuations and hold records on a building by building basis	Feed results back into annual statistics on compliance
	Interface with the maintenance team to log the shower head cleansing on the EMS		Ensure full PPM on showerhead cleaning is completed	Ensure that results are fed back into a compliance register

5. Promote equality actively, support diversity and cohesion and tackle discrimination through innovative strategies	Report on contractor vetting, CRB checks, etc. via the EMS link	Log on EMS system
6. Develop a clear and open communication and engagement strategy	Ensure the needs of the department are clearly demonstrated to the staff through staff meetings	Undertake at least one monthly meeting
	Ensure that all environmental staff get at least a termly one-to-one meeting	Meeting once per term
7. Provide a range of learning opportunities and services relevant to the needs of individuals and employers	Create an EMS link to the training records	Ensure that all training is monitored, especially compliance training, via EMS
	Provide information relating to recycling for the use of students	Attend termly student meetings to pass on recycling progress

(*Continued*)

Appendix IE (*Continued*)

Strategic objective	Action	Department	Target 1	Target 2
8. Provide a supportive environment for all staff encouraging personal well-being and professional development	Provide an end of term report to be issued to all staff on progress made by department		Ensure that progress is fully reported to the whole team	
	Encourage team events		Pull together a set of team events for the year, e.g. team barbeque, liaise with other teams to ensure full involvement	Try to make the team event something to do with recycling
9. Deliver sustainable education; reduce the resource intensity of learning, with the long-term objective to have a carbon neutral site	Ensure a maximum recycling strategy		Report on a monthly basis via service provider on totals of waste and recycled material on site	Make reporting implicit in any new contracts
	Move towards goal of paperless EMS		Have all EMS forms as live versions by end of year	
	Ensure that a diverse range of recycling facilities is made available to the college and all staff		Formal review needed each year	Try new recycling techniques
	Publish data and details of recycling strategy		Publish details of all recycling activities once per year	

Appendix IF Cleaning

Strategic objective	Action	Target 1	Target 2
1. Provide a vibrant learning environment to meet the needs and expectations of the college community	Ensure all areas are maintained clean and serviceable	Undertake a monthly customer care report, personally visiting four customers per month	Achieve a 90% success rate on customer care reports
2. Ensure accountability and financial stability	Cleaning section to undertake internal audits of all cleaning functions and report to Facilities Manager	Ensure that all cleaning activities are reported and a league table of acceptability is drawn up as competition for each building	Create a prize for the staff of individual buildings who perform the best by the end of the year
	Undertake return to work interviews for each member of staff in team	Undertake and keep a formal record of all staff sickness	Report on a termly basis to Facilities Manager
4. Take action to ensure a safe environment. Promote individual understanding and responsibility for safety and well-being	Ensure all toilets/shower areas and places of rest and relaxation are maintained clean and in a condition conducive to good health	Create a schedule of checks detailing cleaning regimes which should feed into the EMS	
	Ensure all cleaning tasks are undertaken in a safe fashion and that records are maintained	Periodic checks on safety must be undertaken by the cleaning supervisors. These must be noted and details provided during monthly FM meetings	

(Continued)

Appendix IF (*Continued*)

Strategic objective	Action	Target 1	Target 2
5. Promote equality actively, support diversity and cohesion and tackle discrimination through innovative strategies	Ensure all contractors are vetted to ensure CRB checks are undertaken	Ensure all staff are fully CRB checked in year 1 of contract	
6. Develop a clear and open communication and engagement strategy	Ensure a meeting structure is in place to provide effective dissemination of information to all staff	Detail and publish a list of cleaning meetings. These should be Cleaning Manager with Supervisors, and Supervisors with operations staff	
7. Provide a range of learning opportunities and services relevant to the needs of individuals and employers	Provide staff training in specialist cleaning in the form of toolbox talks	Annual talks on safety and process should take place and these need to be detailed	Encourage full participation
	Allow succession by encouraging non-paid supervisors	Encourage succession through appraisal to find staff who would like to trade up when supervisors are ill or on holiday	

8. Provide a supportive environment for all staff encouraging personal well-being and professional development	Set up training records for all staff	Set up a competency record for all staff in year 1	
	Provide a non-intimidating, customer-facing service	Cleaning Manager to set up meeting with resource managers in each centre on a termly basis and feedback the results of those meetings to the FM meetings	
9. Deliver sustainable education; reduce the resource intensity of learning, with the long-term objective to have a carbon neutral site	Ensure that all staff use the correct bags for recycling and other waste streams	Termly talks needed to all cleaning staff to ensure the correct bags are used for recycling	Set targets for cleaning staff for recycling
	Ensure a maximum recycling strategy for the college	Periodically reinforce the message on dry mixed recycling with all staff during staff meetings	
	Cleaning section to assist in sustainability by turning off lights and PCs	All staff to be told to make sure windows are closed, PCs are turned off and thermostats are in mid-positions	Reinforce monthly message about turning PCs off

Appendix IG Security

Strategic objective	Action	Target 1	Target 2
1. Provide a vibrant learning environment to meet the needs and expectations of the college community	Provide talks to student groups explaining the role of Security	Annual talks given to students as part of induction process	Obtain feedback from students on security issues
2. Ensure accountability and financial stability	Ensure timesheets are accurate and that all additional hours claimed are ratified by the Facilities Manager before deployment	Ensure timesheet accuracy before submission to Head Office	
	Ensure full complement of staff is maintained	Advise FM office of holidays and ensure full staff requirements	Ensure only trained replacement staff present on site
	Security team to ensure full and timely reports are maintained and followed up with appropriate members of staff	Ensure all incidents are reported	Produce statistics on incidents on a monthly basis
	Undertake return to work interviews for each member of staff in team	Undertake and keep a formal record of all staff sickness	Report on a termly basis to Facilities Manager
3. Set high expectations to promote the consistent achievement of aspirational learning goals	Ensure support to curriculum centres to maintain a safe and non- intimidating level of service is maintained	Discuss with the resource managers on a termly basis the requirements of individual centres, and bring these requests back to the FM meetings	

4. Take action to ensure a safe environment. Promote individual understanding and responsibility for safety and well-being	Support cash-taking activities on site and provide assistance in movement of cash	Provide patrols to ensure security of cash movements	
	Support duty heads during each evening	Provide a sign-on/sign-off facility for duty heads along with a periodic check on status as evening progresses	Gauge effectiveness of service through customer satisfaction surveys
	Undertaken by patrols for safeguarding and security	Ensure red, amber, green patrols are undertaken on a daily basis	Log each patrol and provide statistics on success
5. Promote equality actively, support diversity and cohesion and tackle discrimination through innovative strategies	Ensure patrols are in place to make site feel safer	Report any issues in relation to safeguarding	Provide friendly customer-facing service
	Undertake reports to indicate safeguarding initiatives are working	Provide statistics each month at the security meeting to indicate trends	Compare statistics against previous years
6. Develop a clear and open communication and engagement strategy	A full engagement strategy should be put in place, monitored against targets and reported back through appropriate channels	Discuss with individual curriculum centres the need for attendance at student meetings; draw up a full list and report back through security meetings	

(*Continued*)

Appendix IG (*Continued*)

Strategic objective	Action	Target 1	Target 2
7. Provide a range of learning opportunities and services relevant to the needs of individuals and employers	Assist curriculum centres in the provision of security services	See how the security team can assist via termly meetings with resource managers	
8. Provide a supportive environment for all staff encouraging personal well-being and professional development	Provide a customer-facing service in a non-intimating way	Ensure the service is friendly in its presentation and that periodic customer service surveys are undertaken at least once per year	
9. Deliver sustainable education; reduce the resource intensity of learning, with the long-term objective to have a carbon neutral site	Provide and maintain the eco-secure initiative to check and patrol which buildings are over-using utility resources	The eco-secure system should be reported back to the Facilities Manager at each security meeting and global statistics produced on an annual basis indicating where the hotspots are	Security to report any lights left on

Appendix 2
Management strategy for heating and ventilation in a college environment*

Scope

This document forms part of a working strategy defining how the organisation will approach heating and cooling and when it will apply heating and cooling controls.

The College will aim to hit the CIBSE design space temperatures and will use the CIBSE guidance on heating and cooling.

Building Management System

The BMS controls all the heating and ventilation plant on the site and will turn the heating and ventilation on and off in accordance with the organisation's usage set-ups.

The BMS will be reset each year and changes will be made to take into account the number of occupants and the use of the building.

User surveys

The FM department will survey the building users at least once per term to make judgements on how comfortable the building is. The findings of the surveys will be used to influence and adjust the controls of the BMS.

---

* Adapted from heating, ventilating, air conditioning and refrigeration (CIBSE Guide B), 2005.

Temperatures

The organisation is under a general duty to ensure the health, safety and welfare of employees, so far as is reasonably practicable. This requirement extends to protecting employees and others from temperature extremes.

Low temperatures

The Workplace (Health, Safety and Welfare) Regulations 1992 (further information can be found at: http://www.hse.gov.uk/pubns/indg244.pdf) stipulate that, during working hours, temperatures in workplaces (including maintained and independent schools and FE colleges) should be reasonable. The employer must provide a suitable number of thermometers to enable the temperature to be checked throughout the workplace.

As far as maintained schools in England and Wales are concerned, the Education (School Premises) Regulations 1999 (further information can be found at: http://www.legislation.gov.uk/uksi/1999/2/contents/made) prescribe minimum standards for temperatures (which should be measured at 0.5 m above floor level) as follows:

Use of area	Minimum temperature (°C)	Target maximum temperature (°C)
Lower than normal level of physical activity, e.g. sick rooms	21	24
Normal level of physical activity, e.g. classrooms and libraries	18	24
Higher than normal levels of physical activity, e.g. gyms and drama workshops	15	24

High temperatures

There is no prescribed maximum temperature for educational establishments or any other workplace. The Workplace (Health, Safety and Welfare) Regulations 1992 require employers to take all reasonable steps to ensure that a 'comfortable' temperature is maintained. However, the World Health Organization (http://www.atl.org.uk/health-and-safety/work-environment/temperature.asp) says that, in temperate climates, the optimum indoor temperature is between 18 and 24°C.

Time schedules

Each curriculum centre will submit at the beginning of each year details of opening and closing times for each building area. Wherever possible the individual areas within each of the buildings will be controlled. These time schedules will be uploaded onto the BMS. These will form the core hours from which the heating schedules will be set. From time to time additional hours will need to be provided to cover events such as open evenings. These will be uploaded as special events.

At the beginning of each term the Facilities Manager will request updated information from each centre to ensure that the schedules are up to date.

Thermostatic radiator valves

All TRVs will be set to the mid-position (generally position 3). This is deemed to be the system design setting. Cleaning staff will be instructed to turn all TRVs to position 3 at the beginning of each shift.

Shutdown procedures

To maximise the potential electricity savings, appliances and systems which consume electricity will be turned off as soon as is practical.

At the start of each break site supervisors will be requested to:

- Shut down any PCs that have been left on and turn off at the plug socket. If PCs are switched off, but left on at the socket, they will continue to draw power via vampire power or via the network card. All equipment needs to be switched off at the socket.
- Turn off any projectors and isolate at plug sockets.
- Turn off all printers, photocopiers or ancillary items that have been left on.
- Turn off any zip or point-of-use boilers.
- Switch off or isolate/unplug any supplementary heating.
- Switch off at the socket any ancillary items not required such as kettles, toasters or microwave ovens.

At the end of each term the heating engineer/plumber will be requested to:

- Amend/review any BMS settings and, if necessary, turn off heating or put into setback mode to protect the building from frost damage.

- Check and isolate banks of boilers that will not be needed during shutdown.
- Isolate VRF and/or cooling systems not required during shutdown.
- Amend kitchen ventilation systems.

Considerations for use, design and maintenance

The organisation will endeavour to follow the following principles in terms of use, design and maintenance of all the heating and ventilation plant on site and will include within its PPM details of maintenance that will support the strategies as laid out below.

Space heating (plant)

Use

Support occupants to make effective use of heating controls:

- staff and occupants to optimise heating levels and encourage energy-saving behaviour
- college will promote through internal energy awareness campaigns the reduction and optimum use of resources.

Design

Specify the most efficient primary plant to generate heat/hot water:

- select fuels and tariffs that promote efficiency and minimise running costs
- specify boilers with a SEDBUK rating of A (90% or more efficient)
- segregate hot water services generation wherever possible
- size plant with appropriate margins but avoid over-sizing
- specify condensing boilers
- meet base load using the most efficient plant
- promote renewable technologies whenever practical

Consider controls at an early stage in the design:

- include effective controls on primary plant and distribution systems

- incorporate controls based on temperature, time, zones and variable flow based on the requirements of the building
- ensure controls are functional and calibrated

Provide effective occupant controls:

- put controls (e.g. TRVs) in all rooms/areas with separate control of perimeter areas (within 7 m of external walls) and the central zone

Check the design:

- check installed capacity and energy performance against benchmarks and rules of thumb

Investigate the use of renewable energy or low carbon technologies such as biomass boilers, heat pumps, combined heat and power:

- select the distribution system based on the primary heat source (long lag systems or quick response systems) and the requirements of the system

Minimise heat losses from space heating boilers, pipe work and storage:

- locate plant to minimise loss on distribution system
- insulate pipe work, valves, etc. effectively
- insulate distribution pipes, valves and flanges around the boiler area that are not contributing useful heat to work areas
- avoid external runs on pipe work or keep to minimum

Maintenance and PPM

Regularly inspect and reset controls to maintain efficient system operation:

- reset controls on multi-boiler systems over summer (and in mild weather) so that boilers which are not needed are isolated
- reset controls on unused boilers during the summer months so that pilot lights do not switch on
- reset thermostats (reducing temperature set points) in areas that do not require full comfort heating

- correct frost thermostat settings
- reset heating timer controls to match occupancy patterns
- reset heating timer controls for holidays
- reset set point and control bands on heating and cooling units in the same room to avoid simultaneous operation
- reset controls to avoid unnecessary heating in unoccupied areas, but ensure adequate ventilation to prevent moisture damage

Maintain the heating system to keep it operating efficiently:

- carry out inspection to ensure boiler house ventilation is unobstructed
- inspect feed and expansion tanks where automatic make-up filling can indicate system leaks
- carry out boiler combustion efficiency tests
- carry out regular testing of correct operation of all radiator valves
- test that motorised valves and dampers have full travel from open to closed
- test that hot water valves close fully
- carry out regular testing so that boilers and circulating pumps shut off when the thermostats and heating time switches stop calling for heat
- carry out regular maintenance of boilers including cleaning burners, regular testing of combustion efficiency, adjustment of air/fuel ratio, optimisation of boiler efficiency and complete boiler test sheet
- check and maintain BMS sensors and calibrate to ensure smooth performance

Cooling and refrigeration

Use

Support occupants to make effective use of ventilation controls:

- staff and occupants to optimise ventilation levels and encourage energy-saving behaviour
- the college will promote, through internal energy awareness campaigns, the reduction and optimum use of resources

- fit where possible occupancy controls such as infrared sensors to turn off cooling when no one is in the room.

Design

Apply passive techniques to avoid/reduce air conditioning demand:

- night cooling to avoid installation of air conditioning in mechanically or naturally ventilated buildings
- night cooling to reduce daytime cooling loads in air conditioned buildings

Specify efficient cooling/refrigeration equipment:

- specify refrigeration equipment with high efficiency (high COP)
- follow guidelines for typical and good practice performance indicators for conventional systems as shown in CIBSE Guide F Section 8 Tables 8.1 and 8.2
- refrigeration equipment with high coefficient of system performance (COSP) as this takes into account the power used by all compressors and auxiliary plant as well
- use efficient components such as condensers, evaporators, expansion valves, compressors, pumps and fans
- optimise compressors to increase the proportion of time they operate at full load
- avoid low and head pressure control at condensing temperatures
- use higher evaporating temperatures
- avoid rapid cycling of unloading mechanisms
- auxiliaries can be controlled effectively at low loads
- ensure equipment is easy to maintain and manage

Apply 'free' cooling techniques to reduce air conditioning demand:

- use free cooling systems (especially where the cooling demands are high and unrelated to ambient temperature, for example in computer suites and telephone exchanges)
- use load shaving to combine free and mechanical cooling when ambient conditions are favourable

Raise supply temperatures when possible:

- use larger, more efficient heat exchangers for cooling air
- use increased chilled water or supply air volume flow-rates (with increased pipe and duct sizes to maintain the same pump or fan power)
- use increased chilled water flows through heat exchangers (with increased pipe sizes to avoid increasing pumping energy, e.g. using chilled ceilings)
- separate cooling duties that require low temperatures from more general cooling duties that can be achieved with higher temperatures
- use higher set points at times of lower cooling demand

Design the building structure and fit-out to optimise available thermal mass:

- thermal mass in the structure available to reduce diurnal indoor temperature range (by storing daytime heat, usually to be purged via night cooling)

Consider efficient equipment:

- carry out feasibility study of thermal storage
- heat recovery where this is effective
- ensure that all controls are checked and calibrated annually by the service and maintenance company

Minimise internal heat gains from lighting and equipment:

- use office equipment with lower power use (average heat emissions from equipment as given in CIBSE Guide F Section 12 Table 12.5 & 12.6)
- improve power management controls (e.g. turning off lights and personal computers when not in use)
- encourage reduced usage overall by internal promotion and energy awareness campaigns

Design good HVAC controls:

- optimised set-points (e.g. space temperature, air supply temperature, recirculation rate and humidity)

- improved, but not necessarily closer, control of temperatures, flows and humidity

Design good controls:

- ensure chillers and chilled water systems only operate when there is a real cooling demand, if possible interlinked via the BMS system to prevent both working at the same time.

Size plant correctly (avoid over-sizing):

- size cooling (refrigeration) plant correctly
- size domestic hot water boiler plant sized correctly
- size pumping plant sized correctly
- size air-handling fan plant correctly
- size main space heating boiler plant correctly
- ensure chillers are suitably sized as oversized plant can lead to inefficiency

Reduce solar gains by the appropriate design of the building envelope, glazing and shading:

- put shading devices on windows to reduce excessive solar gains
- paint external walls with a reflective light finish to reduce excessive solar gains

Maintenance and PPM

Regularly inspect and reset controls to maintain efficient system operation:

- reset controls on unused ventilation systems during the summer months so that they do not switch on
- reset thermostats (increasing temperature set points) in unused areas
- reset ventilation timer controls to match occupancy patterns
- reset ventilation timer controls for holidays
- reset set points and control bands on heating and cooling units in the same room to avoid simultaneous operation
- reset controls to avoid unnecessary heating in unoccupied areas, but ensure adequate ventilation to prevent moisture damage

- maintain the ventilation system to keep it operating efficiently
- carry out inspections to ensure ventilation equipment is unobstructed
- carry out regular maintenance of ventilation systems including cleaning/replacing of filters
- carry out regular testing of efficiency, adjustment of air/fuel ratio, optimisation of system efficiency
- maintain F gas register.

Appendix 3
Health and safety rules for contractors

Part I Introduction

The Client xxxx shall hereby be identified as the Organisation.

It is the aim of the Organisation to promote high standards of safety and health at work and concern for the environment throughout its operations. The Organisation encourages the active participation of every member of staff in the promotion, achievement and maintenance of the highest practicable standards of accident and pollution prevention.

The Organisation will promote legislative compliance forcefully and will expect all its Contractors to adhere to legislative requirements. It is also our policy to ensure, so far as is reasonably practicable, that persons not in our employment, i.e. visitors and Contractors and those who may be affected by our activities, are not exposed to risks to their health and safety. Equally, Contractors and self-employed persons who carry out work on our premises have a duty to the Organisation to take reasonable care, comply with statutory duties, cooperate with the Organisation and conduct their undertakings in such a way that it will not adversely affect the health and safety of the Organisation's employees or any other persons on site. These undertakings shall also have no adverse effect on the environment.

The Organisation currently aims to achieve ISO 9001 and ISO 14001 and actively seeks to work with like-minded organisations. The Organisation takes its obligations to the environment seriously and asks all its Contractors and suppliers to have a commitment to improving their individual environmental performance.

This advice note is intended as a guide to Contractors on the general precautions for the more commonly encountered hazards and to encourage the fullest cooperation between the Contractor and the Organisation's representative. The contents of this guide are not exhaustive and therefore the Organisation disclaims any responsibility for any matter or subject omitted from this guide.

The Organisation asks all its Contractors to raise any issues relating to compliance directly with the FM Department so that these can incorporated into this document.

Part 2 General contract health and safety requirements

2.1 Monitoring of contractors and subcontractors

To ensure high standards are adopted and maintained, the health, safety and environmental performance of Contractors will be monitored. If performance is deemed to be unsatisfactory, the Contractor will be liable to dismissal from the site and removed from the approved list of Contractors.

Performance will be compared and related to the health and safety and environmental questionnaires and the method statement agreed at the pre-contract stage. Except for emergency work, Contractors will be required to prepare a brief method statement from information supplied by the Organisation at the invitation to tender stage and from assessment of their own plant, equipment, substances and safe working procedures.

2.2 First day of new contracts

Before commencing any contact, the Contractor's senior representative must report to the Organisation's responsible person for a further briefing. This is to ensure that the method of work agreed at the pre-contract stage is fully understood and will be implemented during the life of the contract. Discussions and introductions may include the following.

- The contract boundaries will be defined.
- Safe means of access and a safe place of work will be established.
- The Organisation's facilities available to the Contractor's employees will be identified, e.g. restaurant, drinking water, first aid, plant, services and supplies.
- Relevant information about the Organisation's processes which may affect the contract.

2.3 Legal requirements imposed on contractors

It is a condition of contracts issued by the Organisation that where work is carried out on premises occupied by the Organisation, the Contractor and any Subcontractors under his or her control must comply with statutory health, safety and environmental legislation together with any relevant procedures in force at the premises concerned.

Your attention is particularly drawn, but not limited, to the current statutory provisions and any future revisions. This list is not exhaustive and Contractors are reminded of their obligations to keep current their individual knowledge of health and safety and environmental issues.

2.4 Common law duties

In addition to statutory provisions, a Contractor is bound to take reasonable care to ensure that both his or her own employees and any persons affected by their acts or omissions at work are safe and not exposed to risks to health. Machinery, work equipment and personal protective equipment brought on to site must be kept in good condition, efficient working order and properly maintained. In addition, all equipment should be inspected to ensure it is in a safe condition prior to any work commencing. In some cases this will be covered by statutory inspection and test procedure, but in all cases PPM measures should be taken to avoid risks to health and safety.

2.5 Supervision of contractors

A key element of any contract on the Organisation's premises is that all Contractors will make adequate arrangements to ensure their undertakings are properly supervised to a high standard. Performance will be monitored by the Organisation's staff, particularly quality control, methods of working and standards of supervision.

2.6 Insurance

Every Contractor must ensure that he or she is adequately insured to the level of indemnity required by the contract. This will cover liability arising from injury and third-party liability.

2.7 Use of organisation's equipment

Except where otherwise agreed, Contractors shall provide all their own tools, plant, equipment and personal protective equipment to ensure satisfactory and safe performance of the required work. All equipment should be in a safe and serviceable condition, properly maintained, covered by a current portable appliance test certificate where appropriate, and comply with all current statutory requirements.

2.8 Electricity, gas and compressed air mains

On no account may use be made of the Organisation's electricity, gas or compressed air mains without the written permission of the Organisation's Facilities Manager. Where permission is granted, the method of connection must be approved by the Facilities Manager.

Part 3 Health and safety procedures

3.1 Below ground services

All Contractors should assume the presence of underground services on the Organisation's premises. To avoid danger from buried services, there are three key aspects to safe working:

- obtain service plans
- use suitable locators
- adopt safe digging practices

Drawings and plans for underground services can be obtained from the Facilities Manager. These will indicate approximate locations that must be verified by the use of locators and careful exposure by hand-digging. All new services must conform to BSI colour coding. Trenches greater than 1.2 m in depth should be adequately supported (see also Avoiding Danger from Underground Services. HSE Guidance Series HSG47, 2000).

3.2 Roofwork

No Contractor is allowed on the roof without written permission from the Facilities Manager. Repairs, maintenance and cleaning work on the roof will be subject to a permit to work.

3.3 Permit to work procedures

The Organisation operates a number of written permits to work procedures. When the contracted work is subject to this degree of control, the permit to work procedures must be invoked and you must ensure that you are familiar with their application. The following operations are covered by a permit to work:

- roof work
- electrical work involving high voltage, uninsulated live conductors, work in electrical substations, adjacent live uninsulated overhead conductors
- hot work, using flame-producing equipment, e.g. welding, cutting, burning and grinding operations
- entry into confined spaces
- work on, or in close proximity to, overhead travelling cranes

Permits must be obtained from the Facilities Manager prior to commencement of work.

3.4 Waste disposal

All Contractors are responsible for the safe removal and disposal of their own waste unless other arrangements have been agreed. No waste materials, either solids or liquids, should be deposited in the Organisation's drains. All toxic materials should be disposed of in accordance with current legislation. Should an accidental spillage occur where a substance enters the drains, this must be reported immediately to the Facilities Manager.

3.5 Demolition work

All demolition contracts must be under the supervision of a competent person. Where demolition works are in progress, adequate and sufficient steps should be taken to ensure that all persons within the vicinity are protected against falling debris and the effects of dust that may arise. For further reference see: http://www.hse.gov.uk/pubns/indg411.pdf

3.6 Entry into vessels or confined spaces

Such work is subject to the Organisation's permit to work procedures.

3.7 Control of Substances Hazardous to Health (COSHH)

Any material or substance brought onto the Organisation's site which is likely to affect the health of either the Contractor's employees or the Organisation's employees will be subject to the provisions of the Control of Substances Hazardous to Health (COSHH) Regulations 2004 (S.I. 2004/3386). Not only may you be asked to ensure that you have in your possession the manufacturer's hazard data sheets listing the properties of the substance, but you will also be required to ensure that copies of the relevant COSHH assessments are available for inspection.

3.8 Reporting of accidents

The Organisation investigates all accidents that occur on its premises and we expect maximum cooperation from any Contractors and their employees to establish causation and the remedial actions necessary to prevent recurrence. All accidents must be reported to the Health and Safety Manager and the details recorded in the accident book. The details of any accidents must be entered in the health and safety file if the contract has fallen within the scope of the Construction (Design & Management) Regulations 1994.

Should an accident occur, work must stop immediately, the appropriate emergency action must be taken (e.g. a first-aider or an ambulance summoned) and the Organisation's Responsible Person informed immediately. The recording of all accidents is a statutory requirement of the Reporting of Injuries, Diseases and Dangerous Occurrences Regulations 2012 (S.I. 2012/199), therefore compliance with this procedure is mandatory.

3.9 Reporting of incidents or dangerous occurrences

A Contractor must at all times report to the Organisation's Responsible Person any incident or dangerous occurrence to enable the appropriate preventative action to be undertaken. A dangerous occurrence is defined in Schedule 1, Part 1 of the Reporting of Injuries, Diseases and Dangerous Occurrence Regulations 1995.

3.10 Fire procedures

Routine fire drills and alarm sounder tests are carried out. When the sounders are being tested, notice will be given in advance. When the alarm sounds at any time other than an informed test, you must immediately make your way outside to the front of the building using the nearest safe exit. Please ensure that you are familiar with your nearest fire exits.

All areas are fitted with break glass alarm points. Make yourself aware of their locations. Should you discover a fire, sound the alarm immediately and the leave the building by the nearest safe exit. When you leave the building, close all doors behind you and move quickly, but without running. Do not stop to collect personal belongings. Never use the lift.

The assembly area is in the visitor car parking area outside the main entrance. Report to your Organisation's Responsible Person and wait at the assembly point until otherwise instructed by this person or a member of the emergency services.

Fire extinguishers are widely available throughout the site.

3.11 Window cleaning

All ladders and other equipment used by Contractors must be of sound construction, adequate strength, sufficient length and properly maintained. Where it is not possible to clean windows from a ladder or where there are any special dangers or risks, the Contractor shall assess the situation and produce a safe working procedure. The Contractor is responsible for checking equipment prior to use.

3.12 Asbestos

No work involving the installation, removal or disturbance of asbestos or work on any article containing asbestos shall be undertaken unless express written permission has be obtained from the Facilities Manager. Any work carried out on asbestos shall be subject to the appropriate legal provisions. If in doubt, please advise the Facilities Manager. A full copy of the current asbestos register is available to view from the Facilities Manager.

You will be informed if your working area contains asbestos or asbestos-type materials. It must be noted that although a full register is held, the Contractor must not assume that all asbestos has been identified. If asbestos-type materials are located, the Contractor must stop work and seek confirmation from the Facilities Manager.

Part 4 Equipment

4.1 Cranes, hoists and lifting tackle

The Organisation's cranes, hoists, lifting tackle and aerial platforms must not be used without prior permission from the Facilities Manager.

Any lifting machine or lifting tackle brought onto the Organisation's site must have a current inspection and test certificate signed by a competent person. These certificates may be requested for inspection at any time. All personnel who are to use this equipment must be fully trained. Proof of training may also be requested at any time. All lifting operations must be carried within the provisions of the Lifting Operations and Lifting Equipment Regulations 1998 (S.I. 1998/2307) and subsequent amendments.

4.2 Ladders, scaffolding and lifting platforms

Where work cannot be safely undertaken from the ground or from part of a building or other permanent structure, provision should be made for safe means of access and safe place of work by means of either scaffold (permanent or portable), ladders (where appropriate) or other means of support such as aerial platforms. In all cases, they must be suitable and sufficient for the purpose and properly positioned and maintained.

Ladders must be pitched at the correct angle of 75°, i.e. 1 m out from the wall for every 4 m of rise, and the top of the ladder should extend five rungs above the landing point (further information is available at: http://www.hse.gov.uk/pubns/indg405.pdf). Outwards and sideways movement should be prevented by securing the ladder at its upper resting place and at the bottom. See Health & Safety Executive Guidance Note GS 31. Tower scaffolds must be properly contracted in accordance with the manufacturer's instructions.

Lightweight DIY-type equipment is not suitable and will not be allowed on the Organisation's site. It should comply fully with the Portable Aluminium Scaffolding Manufacturers Association (PASMA) Code of Practice and the Health and Safety Executives Guidance Note GS 4.

4.3 Cartridge-operated tools

No cartridge-operated tool may be used on the site without consultation with, and agreement of, the Facilities Manager. The intention to use cartridge-operated tools must be disclosed at the pre-contract stage.

4.4 Machinery

All machinery used by Contractors must conform to legal requirements such as the Provision and Use of Work Equipment Regulations 1998 (S.I. 1998/2306). Guards or other protective devices must not be removed from any machine, lift or other plant or equipment provided under agreement with the Organisation. All personnel using the machinery must be trained in its use and may be asked to produce any relevant training records.

4.5 Pressure vessels

Any pressure vessel brought on to the site must conform to the Pressure Systems and Transportable Gas Containers Regulations 2000 (S.I. 2000/128). An up-to-date test certificate should be kept available for inspection at all times.

4.6 Ionising radiation

Equipment and sources, either sealed or unsealed, which produce ionising radiation subject to the Ionising Radiation Regulations 1999 (S.I. 1999/3232) must not be brought on to the site without authorisation. Radiation hazards need strict control and disciplines, which must be maintained. If the contract requires work to be undertaken in these areas, the Contractor will be subject to strict control procedures.

4.7 Other protective equipment

It is the Contractor's responsibility to ensure that adequate and suitable safety barriers, lights, safety signs and hazard warnings are provided and used.

4.8 Electrical equipment

No unauthorised person is permitted to adjust, repair, connect, disconnect or tamper with any part of an electrical installation, switchboard, wiring, fittings or portable tools. Any defects must be reported immediately to supervisors. Portable tools (110 V) must be used unless written permission has been obtained from the Facilities Manager.

4.9 Hand tools

All hand tools brought on to the Organisation's premises must be properly maintained and the correct tool used for the job. Improvised equipment and tools must not be used. All portable electrical appliances must have a current test certificate, which must be made available for inspection if requested.

4.10 Gas cylinders

All persons working with gas cylinders must be fully acquainted with the dangers associated with the various types of gases stored in the cylinders. Gas escapes may result in the risk of fire, explosion and/or the inhalation of toxic gases. They must be transported correctly in open vehicles and stored securely in well-ventilated conditions out of direct sunlight.

Part 5 People

5.1 Personal protective equipment

It is the duty of the Contractor to ensure that employees and any persons under the control of the Contractor are supplied with and use appropriate personal protective equipment and that the wearers are trained in its use. Provision and use of personal protective equipment must be in accordance with the Personal Protective Equipment at Work Regulations 1992 (S.I. 1992/2966).

5.2 Contractor's vehicles

It is forbidden to take any vehicles into a building or structure without prior permission from the Facilities Manager. All Contractors must comply with the Organisation's traffic management procedures. Drivers must at all times exercise utmost diligence as this site has a high volume of pedestrian traffic. All traffic accidents or near misses must be reported to the Health and Safety Manager immediately.

All drivers of vehicles must have undertaken relevant training and medical examinations. Proof of this may be requested at any time. All transportation involving vehicles across pedestrian areas can only be undertaken with the use of a banksman. Contractor parking on site is limited.

5.3 Contractor's personnel

All personnel provided by Contractors to carry out work must be adequately trained, competent and fit to carry out the work required. Records of such training may be requested for inspection at any time. Contractors will be expected to behave in a manner commensurate with all precautions necessary for work in the vicinity of vulnerable adults. Safeguarding is of paramount importance and the Organisation reserves the right to remove from site any member of the Contractor's staff whom it believes is not complying with current safeguarding legislation.

5.4 Personal hazards

Running, horseplay or the playing of games in or near buildings, property, roadways or car parks is strictly prohibited, as is the wearing of personal stereos. Offenders may be removed from site.

5.5 Subcontractors

The Contractor shall at all times be responsible for the actions of any Subcontractors and ensure that they work to the Organisation's health and safety rules and regulations.

5.6 Alcohol and drugs

Alcoholic drinks or drugs should not be brought or consumed on the Organisation's site. Should drugs be required for particular medical conditions, the Organisation should be notified prior to work starting. Any individual suspected of being under the influence of drugs or alcohol will not be allowed to continue work on the Organisation's site.

Part 6 Receipt and declaration

I have read and understood the Organisation's Health and Safety Rules for Contractors and agree to abide by these rules at all times whilst on the Organisation's site.

Furthermore I/we confirm our commitment to improving our environmental performance. I/we also agree to ensure any wastes produced while on site are removed and wherever possible recycled.

Name (Block Capitals):

Company (Block Capitals):

Signature:

Date:

Appendix 4
Sustainability policy

This Organisation is committed to minimising the environmental impact of its activities through adopting sound principles and implementing good practice, and by continuous improvement in environmental performance. The Organisation will seek to develop an ethos of environmental responsibility and reduction of the resource intensity in all its staff and students as well as in the management of its estate and, in doing so, aim to become a sustainable community.

Core policy statement

This Sustainability Policy will act as a guiding document for all future improvements and is linked to all other Organisation policies. Accordingly, it is the policy of the Organisation to:

1. Manage the Organisation in a manner that reduces the resource intensity and impact, with emphasis on reducing its carbon footprint.
2. Seek cultural change through raising awareness of collective and individual responsibility in working towards achieving a sustainable community.
3. Continually increase its knowledge base of sustainability issues.
4. Consider the wider environmental impact of its spending decisions through seeking an appropriate balance between cost and sustainability and apply sustainability criteria to procurement.
5. Conform to and, where possible, exceed minimum legal requirements for reducing adverse environmental impacts.
6. Work with partners and external communities to help promote sustainability within the local area.

7. Monitor and regularly audit its environmental performance and develop management systems that seek continuous, measurable progress and report this to the directors.
8. Ensure that this policy and all reports and reviews of environmental performance are communicated to all staff and are available to the public.

Sustainability targets

The policy will be monitored by the Estates and Facilities Manager, who will also be responsible for its implementation. The Organisation will seek to achieve the following through its Carbon Reduction Plan:

1. Develop targets for a reduction in energy use, including the adoption of conservation measures and renewable energy sources.
2. A reduction in water usage through wise use and adoption of conservation measures.
3. A commitment to the wise choice of products and suppliers and the use of sustainable procurement processes.
4. Minimisation of waste through sensible purchasing, reduced consumption and the adoption of efficient waste management and recycling methods.
5. Improving the quality of life on the site through encouraging appropriate and considerate behaviour in relation to litter and antisocial behaviour.
6. Adoption of contingencies to reduce the environmental impact of transport/travel and to improve accessibility for all.
7. Adoption of best practice in new build and refurbishment projects, as set down by BREEAM assessments (http://www.breeam.org/page.jsp?id=301).
8. Improved communication between all sections of the Organisation to promote shared responsibility, to celebrate achievement and success and to further improve the quality of life at the Organisation for all.
9. Inclusion of staff training/CPD covering sustainability issues and coverage of policy and practice in staff induction.
10. To increase the use of renewable technologies on site

Carbon Reduction Plan

The Carbon Reduction Plan will be a live document that prioritises actions to achieve greater sustainability at the Organisation over the short and long term. It will: identify its area of responsibility; identify key players and responsibilities; identify key targets, both long and short term; and monitor progress against targets.

Appendix 5
Dealing with bomb threats

Bomb threats may be made to the Organisation; however, careful management of the situation and adherence to defined procedures will minimise the risk to staff, visitors and property. Any intimation, no matter how vague, that any explosive device has been planted must be regarded as a threat until investigation and subsequent events prove otherwise.

Bomb threats are usually received by telephone and most are hoaxes, designed to cause the maximum disruption. However, all should be treated initially as though they are genuine; the making of such calls is a crime and they should always be reported to the police.

The following information is adapted from Home Office (1994).

Immediate action to be taken on receiving a telephone bomb threat

All switchboard operators, receptionists and staff who receive a telephone bomb threat must keep calm, be courteous, listen and carry out the following procedures.

1. Allow the caller to finish the message without interruption.
2. Ask the caller:
   a. Where is the bomb right now?
   b. When is it going to explode?
   c. What does it look like?
   d. What kind of bomb is it?
   e. What will cause it to explode?
   f. Did you place the bomb?
   g. Why?
   h. Who are you?
   i. Where are you?

3. Complete the Bomb Threat Checklist as fully and as soon as possible after the call.
4. Immediately attempt to trace the number by dialling 1471.
5. If the call is received via the switchboard or a phone that shows the telephone number on an LCD screen, try to make a note of the number before the caller rings off.
6. Notify a member of the Executive, Senior Management Team or Duty Manager (in the evening) giving as much detail as possible.
7. The member of the Executive, Senior Management Team or Duty Manager will immediately notify the police using the 999 emergency number. WARNING: mobile phones must not be used to make calls as this could trigger a bomb. The Senior Manager who takes charge will liaise with both the police and the Organisation's executive staff.

An evacuation of the building and adjacent buildings within 100 m will only take place if advised to do so. An evacuation must be silent and not via the fire alarm system as the sounding of fire alarms trigger a bomb. A search of the building will carried out only if asked by the police to do so.

Bomb searches

Responsibility for the search of premises rests with the Organisation with assistance from the police. It must be remembered, however, that the police will be unfamiliar with both buildings and contents. If, after discussion with the police, the Senior Manager in charge determines that a search is required, the following points should be noted.

- A bomb may take many forms. It is the unusual object not normally found in that location which is suspect.
- Searchers must be methodical, with each searcher being responsible for predefined areas.
- Prime bomb planting sites include lobbies, cloakrooms, toilets and car parks.
- Any suspicious object must be interfered with. The police must be advised and any subsequent actions be carried out under their direction.

Receipt of a letter or fax containing a bomb threat

The person receiving the threat must immediately alert a member of the Executive or Senior Management Team. The member of the Executive or Senior Management Team will immediately notify the police using the 999 emergency number.

Suspect or contaminated packages and letter bombs

If a suspect package is found, then the person discovering the package must ensure the following procedure is carried out:

- Do not touch the package. Inform a member of the Executive or Senior Management Team who will notify the police immediately using the 999 system. Do not use a mobile phone.
- Shut windows and doors in the room, if possible, and leave the room. If it is suspected that the package may be biologically contaminated, it must be kept separated from staff, who must be available for a medical examination.
- Do not switch electrical equipment on or off.
- If possible, switch off any room air conditioning system. Contact the FM department for an emergency shutdown of the building's ventilation systems via the BMS.
- Manually close all fire doors in the building.
- If there has been a suspected chemical incident, staff are to leave the area immediately. Signs that people may have been exposed to a chemical incident are streaming eyes, coughs and irritated skin. Medical advice should be sought immediately in this instance.

Any one of the following signs should alert staff to the possibility that a letter or a package contains an explosive device.

- Excessive grease marks on the envelope or wrapping.
- An unusual odour, such as marzipan, almonds or machine oil.
- Visible wiring or tin foil, especially if the package is damaged.
- The envelope or package may feel very heavy for its size.
- The package may have been delivered by hand from an unknown source or posted from an unusual place.
- The package may have excessive wrapping.

The member of staff who discovers the device will be required to give the following details for the assistance of police and military teams: size, shape, length, diameter/width (a sketch may help), postmark, any details of sender, whether excessive number of stamps used, any irregularity in address, e.g. spelling errors, wrappings, presence of Sellotape, balance and any other potentially useful information.

In all cases of suspect mail, whether explosive devices or threatening/ hoax letters, police can use the following as evidence: when, where and by what means delivery was made; who accepted the item on behalf of the company; who later had possession of the item; and who finally handed the item to the police.

On no account should any member of staff ever agree to look after packages or containers for people not known to them.

Bomb threat checklist

Circle applicable word below; insert words where necessary.

Time:

Date:

When is bomb due to explode?

Where is bomb placed?

Ask identity of caller/organisation.

Origin of call: STD code/coin box/internal?

Caller: male/female/adult/child?

Voice: loud/soft/rough/refined?

Pitch of voice: high/medium/deep/disguised?

Speech: fast/slow/distinct/blurred/stutter/lisp?

Language: obscene/coarse/normal/educated?

Accent: local/regional/foreign?

Manner: calm/angry/rational/irrational/coherent/incoherent/deliberate/ hysterical/humorous/drunken?

Background noise: factory/road traffic/music/office/party/PA system/animal noises/railway/street noise/quiet/voices/other?

Text of conversation: ???

Appendix 6
Preliminaries and conditions

Preliminary particulars

The Contractor shall be deemed to have allowed against each item for the cost of complying with these Preliminaries. All items contained in the Preliminaries shall apply to the whole of the works.

1 Names of parties

1.1 Employer: Referred to as: the Organisation.

1.2 Scheme: works to xxxx.

2 Design of works

2.1 The Organisation will prepare a specification and schedule of works. The Contractor shall allow in their price for the preparation of setting out drawings necessary to complete the works in addition to site surveys, including the taking of site dimensions and levels.

2.2 Dimensions. No dimensions shall be scaled from drawings. Before any work is started, relevant dimensions shall be taken on site and the Contractor shall be responsible for the accuracy of these dimensions.

2.3 The Contractor shall read this document as an Annex to the specification.

3 Inspection of site

3.1 The Contractor shall visit the site and satisfy themselves as to its accessibility, the nature and content of the works and the circumstances under which the works will be executed, as no claim on the grounds of want of knowledge in this respect will be considered.

3.2 Permission to visit the site shall be obtained from the Facilities Manager.

4 Programme

4.1 The Contractor will be required to submit for approval a fully detailed programme, clearly indicating the method proposed to carry out the works and the time each event is expected to take. The programme should incorporate direct and nominated Subcontractors' work and the Organisation's own directly employed Contractors, which, when agreed, shall be strictly adhered to.

5 Tender

5.1 The Contractor will be required to submit a fixed-price tender based on a specification, schedule of works, contract period and any other conditions contained within the contract documents. They will be asked to demonstrate within the contract price compliance with CDM Regulations if needed.

6 Contract particulars

6.1 The agreement and conditions of contract shall be those issued by the Joint Contracts Tribunal for Minor Building Works (Sweet & Maxwell: 2011) or standard letter specification.

6.2 The specification will form part of the contract.

6.3 Codes of Practice. The Contractor is referred to the various clauses in the Conditions of Contract, together with any alterations and amendments not included subsequently in this bill, and of which the following are short references, and they are to provide here any sums considered necessary to cover them.

6.4 The contract period will be from xxxx to xxxx.

6.5 Insurance

The Contractor will be held responsible for, and is to indemnify the employer against, all claims in connection with noise, vibrations, dust or any nuisance arising from the execution of the works. The employer will insure the existing works against loss and damage by fire, etc. Allow for insuring all Contractors' and Subcontractors' plant, tools and equipment, including site offices and storage sheds, against loss or damage howsoever caused. The Contractor will be required to furnish the Organisation with details of insurance's relevant to the contract, thus:

Name of insurer:

Sums/extent of insurance:

Date and serial number of last premium payment:

6.6 Plant, tools and vehicles. Allow for providing all plant, tools and equipment and vehicles necessary for the proper execution of the works.

6.7 Safety, health and welfare of workpeople. Allow for all costs incurred in complying with all statutory and working rule agreements and requirements for the safety, health and welfare of workpeople and staff. The contractor will be expected to provide (if requested) full risk assessments for the execution of the works.

6.8 Overtime. The Contractor shall include all labour costs for carrying out the works, including all extra costs of non-productive overtime and the like necessary to complete the works within the contract period. Allow for the payment of National Insurance and pensions for workpeople and any other contributions, levies and taxes payable by the Contractor in his capacity as an employer.

6.9 Holidays. Allow for the payment of all charges in connection with holidays with payment agreements.

6.10 Staff. Allow for all costs incurred in transporting workpeople and staff to the site. The Contractor shall ensure that their staff carry out their duties while on the Organisation's premises in an orderly, safe and courteous fashion, given the nature of the services they are providing. The Contractor may be requested to ensure staff are CRB checked or work under constant supervision. In respect of all persons employed or seeking to be employed by the Contractor, the Contactor shall comply with each and every one of the provisions in law which prohibit discrimination in relation to employment on the grounds of sex, colour, race, disability, ethnic or national origin or religion. The contractor shall ensure that all staff under their employment maintain an acceptable level of personal hygiene. Contractors are requested to ensure that all staff wear clothing with badges which indicate the name of the contractor. The Contractor shall ensure at all times that appropriate personal protective equipment is provided for use by their staff. The Organisation may reasonably request, that certain members of the Contractor's staff be removed from the premises if in their opinion they represent a threat to other staff.

6.11 Safeguarding against damage or theft. Allow for safeguarding the works, materials and plant against damage and theft, including all necessary watching and lighting for the security of the works and the protection of the public.

6.12 Maintenance of roads. Allow for indemnifying the employer against all claims from the Local Authority in respect of damage to public roads and paths caused by traffic to and from the site in consequence of the works.

6.13 Allow for keeping roads and paths immediately adjacent to the site free from mud, dust, rubbish and obstructions arising from the execution of the works, and for complying with local bye-laws or regulations relating to the cleanliness of streets and public rights of way.

6.14 Police regulations. Allow for complying with all police regulations affecting the site or the execution of the works.

6.15 Materials and workmanship. All materials shall be of the best quality consistent with the character of the works, generally in accordance with the latest British Standard specification where applicable, and to the complete satisfaction of the employer. The workmanship shall be of a high standard throughout, particularly with regard to the accuracy of dimensions, lines, planes and levels, and to the quality of finishing and to the complete satisfaction of the employer.

6.16 The Services. The Contractor shall during the contract period allow for four meetings on site. These visits will be coordinated with the Organisation and will be adhered to. The contract shall have in place such resources as to deal with emergency breakdowns out of hours and will respond to such breakdowns as an emergency (within two hours). Any amendments to the service provided will only be after receipt of written instruction from the Organisation. All work shall be carried out in accordance with: all relevant Acts of Parliament, Statutory Instruments and Regulations; any public health, security and contract requirements as may from time to time be issued by the Organisation; any relevant safety regulations published by the Organisation. All work and materials shall comply with relevant British Standards and Codes of Practice.

6.17 Samples. Allow for providing samples of the various materials or goods from time to time as required. The Facilities Manager will retain samples and any goods delivered which are not up to the standard of the samples will be rejected.

6.18 Water for works. Allow for ascertaining the source of supply and for providing water for the works, and paying all fees and charges in connection therewith.

6.19 Temporary lighting and power. Allow for ascertaining the source of supply and for providing lighting and power for the works, and paying all fees and charges in connection therewith.

6.20 Locating and maintaining existing services. Allow for taking all reasonable measures to ascertain the location of all existing services, including liaising with the local authorities, public undertakings, the employer's representative and any other persons having knowledge of the site, to avoid damage to these services.

6.21 Allow for protecting, upholding and maintaining all existing services during progress of the works, and for giving all notices to the statutory authorities. The Contractor shall make good any damage due to any cause within their control at their own expense or pay any costs and charges in connection therewith.

6.22 Temporary works: general. The siting of temporary hoarding and screens shall be agreed with the Facilities Manager before work commences and the Contractor shall allow for removing the same to other agreed positions to ensure the proper progress of the contract.

6.23 Allow for obtaining permission and for complying with the requirements of statutory and local and other authorities in respect of temporary hoarding, buildings, screens, gantries or scaffolding and pay all costs attached thereto.

6.24 Temporary hoarding and gantries. Allow for providing temporary hoarding, planked footways, guard rails, warning lights and the like for sectional possession of the site, for protecting the public for the proper execution of the works and for meeting statutory and any local authority requirements. Footpaths to be closed off and watched while overhead work is being carried out.

6.25 Since the works are to be carried out on the Organisation's property, no space can be allocated for the storage of Contractors' materials and plant other than within the curtilage of the Organisation's buildings.

6.26 No placards or advertising shall be fixed to any temporary hoarding, gantry, scaffolding or building structure without the employer's express approval.

6.27 The right of advertisement on hoarding, gantries, etc. and on the building is reserved to the employer absolutely.

6.28 Allow for obtaining all licences and paying all fees and charges in connection with the foregoing works.

6.29 Scaffolding. Allow for providing all scaffolding, including special scaffolding, where required for the proper execution and completion of the works, including altering, shifting and clearing away, and making good on completion. Ensure scaffolding is provided over entrance doors to protect the Organisation's staff and visitors.

6.30 Permission of adjoining owners. Permission to be obtained from adjoining owners and occupiers before erecting any hoarding, gantries, scaffolding, etc. on, above or overhanging their property.

6.31 Works by public bodies. Sums for works to be carried out by a local authority or public undertaking will be net, and no allowance will be made by any local authority or public undertaking in respect of cash or any other discount.

6.32 The term 'general attendance' shall be described for nominated Subcontractors.

6.33 Work by nominated Subcontractors. General attendance on nominated Subcontractors shall be allowed in the tender price. The term "general attendance" shall be deemed to include arranging with nominated Subcontractors the time for commencement of their work on site, obtaining from them particulars of holes, mortices, chases, recesses, bases and supports for equipment, fixings and the like, and supplying them with all dimensions and other information required for the proper execution of their works. Where applicable, the Contractor shall obtain the employer's approval to Subcontractors or suppliers setting out and/or shop drawings before the work is put in hand.

6.34 Invoices, receipts and vouchers. The Contractor shall produce for inspection, when required, all invoices, receipts and vouchers in respect of payments for items for which price cost sums and provisional sums are provided. The contractor will provide a schedule of dates to be invoiced which will tie into service visits. Invoiced amounts will be for works completed only.

6.35 Security. The Contractor shall be responsible for taking all reasonable care that the property is secure from vandalism, theft and materials taken from site. In particular, the Contractor will be responsible for ensuring that no unauthorised persons are on site during working hours and that all gates, doors and windows are securely locked and bolted. All work involving the opening up and providing access into the building must be properly secured out of working hours and weekends to prevent any entry.

6.36 Protecting from damage. The Contractor shall protect the works from damage and allow for making good any damage or loss arising from insufficient protection or damage to previously completed work, or from carelessness of the Contractor's workers. The whole of the works including those of Subcontractors, will be under protection of the Contractor.

6.37 Protecting from the weather. The Contractor shall protect the works from inclement weather and allow for maintaining such protection and making good any damage or loss arising from insufficient protection.

6.38 Drying the works:. The Contractor shall allow for the correct and adequate drying of the works and control of humidity, and shall provide all necessary labour, fuel and appliances to comply with these requirements, and to facilitate trade following trade and the work of nominated Subcontractors.

6.39 Removing rubbish and cleaning. Allow for removing rubbish and debris to maintain a clean and unrestricted site at all times or as may be directed, including that caused by Sub- or other Contractors. All rubbish and debris shall be removed from site at the end of each working day.

6.40 On completion, thoroughly clean all areas of working and areas affected by the works. Leave the site clean and tidy.

6.41 Trade union membership. Allow for complying with all national and local trade union requirements as no claims on grounds of want of knowledge in this respect will be considered.

6.42 Travelling and subsistence allowances. The Contractor is to allow for all costs and payments to work people as arising from their employment on the works.

7 Subcontractors

7.1 The Contractor shall be responsible for the supervision and administration of all Subcontractors.

7.2 The Contractor shall submit for approval, on request, a list of firms proposed for employment as Subcontractors for trade or specialist items.

7.3 The Contractor shall provide exact and legible documents to all firms where it is proposed to sublet any portion of the works.

8 Health and safety of contract workers

8.1 The Organisation acknowledges that it bears some responsibility under the Health and Safety legislation for the safety of employees of other organisations who work temporarily at or visit premises owned by the Organisation. The Organisation will use the guidance as found in the HSE publication 'Use of Contractors' available for free at: http://www.hse.gov.uk/pubns/indg368.pdf

8.2 As far as can be foreseen, there should be no special or unusual danger facing your employees whilst working on the Organisation's properties and a compliance with the Construction (Design and Management) Regulations 2007 (S.I. 2007/320).

8.3 In all matters relating to the work undertaken by your employees, the Organisation will require that you, as the employer, have met your obligations under the Health and Safety at Work Act 1974, and that your employees are aware of your safety policy, their obligations and that they abide by them.

8.4 Contracting or Subcontracting companies' personnel working on the Organisation's premises must at all-time comply with the legal requirements of any regulations currently in force.

8.5 Contractors' or Subcontractors' employees must avoid doing anything likely to endanger themselves or other persons engaged on the Organisation's premises.

8.6 No excavations or work on the Organisation's premises, structure or plant and equipment therein, is allowed until permission is obtained from the Facilities Manager.

8.7 Excavations once commenced must be securely fenced and boarded over if necessary. They must be lit by lamps during darkness. Debris must not be left in a manner in which it presents a hazard. The site must be cleared on completion of the work.

8.8 Contractors' or Subcontractors' personnel must notify adjustments to services to the officer in charge.

8.9 Precautions must be taken in the event of the use of welding or flame cutting equipment. Areas must be screened. The Organisation operates a permit to work scheme and all hot works must comply with a hot work permit. A similar permit to work system operates for all works to roof areas.

8.10 Contractors and Subcontractors must provide their own equipment which should be properly maintained.

8.11 Dangerous occurrences, whether involving injury or not, must be reported to the officer in charge.

8.12 The Organisation's Facilities Manager has the authority to prohibit the use of dangerous equipment, or to order off site anyone working in a manner dangerous to themselves or others.

8.13 All accidents resulting in personal injury must be reported and a brief statement of the accident entered in the Organisation's accident book.

8.14 The Contractor or Subcontractor shall supply adequate and suitable protective clothing and equipment, e.g. safety helmets, goggles, boots, for their workers and ensure that they are used whenever necessary. The Contractor should attempt to engineer out the need for personal protective equipment, which should be supplied as a last option to mitigate risks.

8.15 Contractors and Subcontractors must confine themselves to their area of work and obey all notices.

8.16 The contractor shall restrict hours of work to 8.30 a.m. to 6.30 p.m. Monday to Friday and 8.30 a.m. to 12.30 p.m. Saturdays, except with the express prior permission of the Organisation.

8.17 Contractors must report to the FM office on arrival on site and seek any permits that may be needed for the day's works.

8.18 No access to the site for Contractor's vehicles can take place between 8.00 and 8.55 a.m. and 3.10 to 4.00 p.m.

9 Quality

9.1 The Contractor shall institute and maintain a properly documented system of quality control to ensure that the contract is being executed in a proper fashion in relation to all Regulation, Codes of Practice and general good practice. The Contractor if requested shall make this system available for the Facilities Manager to view. A formal review of the services provided will be undertaken at the end of year 1. The contractor will be expected to show an improvement in the service provided in year 2 of the contract.

10 Contract price

10.1 Subject to the contract particulars in this document and the attached specification, the Organisation shall pay the contractor for the provisions of the services as defined.

10.2 If for any reason the agreement is terminated otherwise than at the end of the month, then without prejudice to any other rights of the Organisation, including set off and counter claim by the Organisation in respect of any partly complete work or month, shall be duly apportioned as at the date the agreement ends or is terminated and the proper proportion paid in accordance with the terms hereof.

10.3 The Contractor shall have deemed to include all employees' rates of pay and potential increases within the tendered figure; any request for uplift will be duly declined.

Note: any users of this information must ensure the legality of any or all of its contents before use.

Appendix 7
Strategy for helpdesk

The organisation operates a Helpdesk, which is defined as the central hub for all FM jobs. It should input, coordinate and monitor all the FM workload.

The Helpdesk will deal with:

- all incoming internal jobs
- liaison with, and booking in of, contractors
- coordination of work required
- action of emergency works
- customer feedback
- service standards and audits

Management responsibility: the Facilities Manager is responsible for this strategy.

Operational responsibility: the Estates Administrator is responsible for the day-to-day running of the strategy.

All incoming jobs

It is essential that all jobs or processes are booked into the Helpdesk software. This enables a full audit of the work undertaken. It will enable full monitoring of the FM workload and creates justification for staffing and workloads.

Liaison with contractors

The Helpdesk needs to be at the centre of the FM function and needs to be able to summon contractors when needed and when the FM team cannot deal with a problem. Contractors coming on site must report to the Helpdesk, who will coordinate the response.

Coordination of work required

The Helpdesk will coordinate any works required in terms of allocation of jobs from initial request and/or any additional works that are necessary as a result of the completed job. For example, if a lift breaks down, the Helpdesk will request assistance from the Site Supervisors to check this out. The Site Supervisors check the lift, but cannot resolve the problem. The Helpdesk then contacts the lift service engineers.

Action of emergency works

These are events that need to be dealt with immediately, such as a flood or electricity outage. The Helpdesk will contact the appropriate members of the FM team by radio to deal with the emergency depending on requirements. This initial radio contact will then be backed up with allocation of a formal job by the Helpdesk.

Customer feedback

The Helpdesk will contact a percentage of the staff who have logged jobs onto the system, either through the web jobs option or via telephone, to assess customer satisfaction. This will be reported via the FM team meetings.

Service standards and audit

The Helpdesk staff will periodically undertake sample audits on the services provided, e.g. cleaning. They will visit work centres and make formal lists of their findings and will forward these findings to supervisors/managers for action. These will be logged onto the system to ensure completeness.

Service audits can be carried out on cleaning, grounds, maintenance, building condition, access and egress and lighting.

Helpdesk jobs process

New job arrives

- Job rung into Helpdesk (all jobs must go through the Helpdesk before being started). If staff stop and ask a member of the FM team to do a job, refer them to the Helpdesk. Do not offer to do the job yourself.

- Job may come in via the web jobs option on the Helpdesk software. Staff enter the details onto the Helpdesk and start the five working day clock ticking.

Job issued

- Supervisors to print out all jobs that have come in the previous day, or that morning for the late shift. This should be carried out once each day for Maintenance and twice each day for the Site Supervisors before each shift. This to be done before the workforce arrives for work.
- Jobs will be placed into individual pigeonholes for staff to pick up.
- Staff will pick up all jobs for the day once, and once only, when they start their shift.
- The only exception to this will be emergency jobs, which will be radioed out to the individual responsible; the paperwork for the job will be in the pigeonhole the next morning.
- If a member of staff is absent, then his or her immediate deputy will be responsible for printing out the jobs.

Job process

- Staff will concentrate on completing the jobs issued that morning unless a planned job or emergency job takes precedence.
- Not all jobs will be completed immediately, e.g. if they require parts. Parts/supplies need to request a requisition form, which will detail the job number, parts required by what date. If parts can be purchased before the five-day expiry then they will be ordered and the job will stay on the Helpdesk. However, if parts are ordered and they cannot be delivered before the due date, the works will be taken off the Helpdesk and go onto the plan board. Feedback to that effect will then go onto the Helpdesk.
- Staff will take the requisition form to the Helpdesk Manager who will price the request. The Helpdesk Manager will then take the priced item to the Maintenance Supervisor to sign off. He or she will check whether the job is valid and whether it is a job FM are paying for or whether another cost centre is paying.
- The Helpdesk Manager procures the items (staff need to allow time to procure the item, so early requests are essential).

Job completion

- It is important that feedback is updated and kept current.
- It is important that jobs are not signed off as complete until they have been completed.
- Once a job is complete, the docket must be returned to your supervisor so that it can be closed down.
- To save paper, jobs should not be printed out multiple times.
- Jobs should be closed down in a batch once each day.

Appendix 8
Strategy for operations and maintenance

The Organisation is committed to good management practices in its administration of the FM Department. In particular, it is committed to holding all the site information on the current O & M strategic indexing system.

Management responsibility: the Facilities Manager is responsible for maintaining this policy and for its content.

Operational responsibility: the Facilities Administrator is responsible for the upkeep of this strategy, and also each of the FM teams.

The principles of the system are:

- to hold all information pertinent to a single building on one file or series of files
- to replicate this process for the whole site
- to hold all information in one central area
- to use a uniform process for each building to assist retrieval of required data
- to manage these data and control entries and omissions
- to control information taken out of the O & M records by formal sign-in, sign-out

The O & M files will be the mainstay for the following information:

- building details, size and age of premises
- building construction information, such as who designed and built it, copies of any permissions, details of commissioning data and copies of instructions and warranties

- details of PPM
- catering information
- contractual advice slips from maintenance contractors
- testing certificates
- miscellaneous information

Supplementary to the O & M Manuals, plans of all the buildings will also be held in paper format. These will be managed in the same way in terms of signing in and signing out.

It is generally understood that as our data storage capability improves, we will upload all existing files onto the computer system to reduce our use of paper. This will help in achieving our sustainability aims.

Appendix 9

Service level agreements

Your organisation has many departments which provide various services. Some of these services are provided to internal clients and some to external clients. The commonality here is the word "service." Providing a service is a fundamental aspect of what we do. Almost our entire workload consists of services we provide. To define a Service Level Agreement (SLA), it is first important to provide a definition of the word service in the context of an SLA.

> A "service" is the process of meeting a requirement, need, request or function. It may form part of a process or be a standalone process. It is defined and specified and needs to be documented and provided to an individual or group. It can be part of a chain of actions that create an outcome. The outcome must be able to be monitored and measured.

An SLA is an agreement or contract with a supplier of a service; the supplier can either be internal or external to the organisation. An SLA defines the service the supplier must provide and the level of service to be delivered. They also set out responsibilities and priorities. SLAs can be contractual obligations and re often built into a contract, where they may form one or more clauses or an entire section. SLAs can be used in any supplier contract where a service is provided. SLAs are complex documents which should be well defined and cannot be drawn up in an ad hoc fashion. Taking the time to integrate SLAs into your overall framework will help you to monitor quality and help set bench-marks and targets.

Typical SLAs are set out as follows.

- the service or services being provided
- the standards of service and the level of quality required
- the timetable for delivery of the service, including key landmark dates

- provisions for legal compliance
- payment terms and possible compensation arrangements
- the responsibilities of suppliers and customers
- mechanisms for monitoring and reporting service and Quality Management
- a series of criteria by which the standards will be measured
- how disputes will be resolved, who will resolve them, how long the resolution will take and who will arbitrate in the event of a stalemate
- confidentiality and non-disclosure provisions and the penalties for breaking these
- termination conditions by both parties

If suppliers fail to meet agreed levels of service, SLAs usually provide for compensation, commonly in the form of rebates on service charges or, on internal contracts, providing a rating to demonstrate the level of service provided/achieved.

When drawing up your SLA, highlight the most critical components of the deal so you can apply the strictest penalties or severest ratings to these elements. These will be the critical deliverables for the contract.

Build periodic performance reviews into the SLA. These can be a standard agenda item at your regular meetings. Review, review and review again. Only when you have looked at all the potential for the service can you be sure you have achieved an outcome based on quality. You may never be satisfied, but this demonstrates a continued improvement in quality.

Appendix 10
Cleaning specification*

This specification is intended to be an output specification and as such only the required output/target is specified. The organisation has not attempted in any way to detail methods, practices or techniques that may be required to achieve the required output/target.

Sanitary/washing appliances

Wash-hand basins/sinks/taps/surrounds/pipes/shower trays

The inside and outside of sinks and shower trays shall be cleaned daily and as necessary throughout the day and left free of stains and marks. The contractor shall ensure that waste gratings are kept free from obstruction. The external surfaces of all sanitary fittings, pipes and supports shall be cleaned with descalar, dried and polished as the finish of the fitting, pipe or support dictates.

China sinks

The internal and external surfaces of china sinks shall be scrubbed and dried daily and as necessary throughout the day to ensure that they are kept free from stains and marks. The contractor shall ensure that waste gratings are kept free from obstruction. The pipes and supports shall be cleaned with descalar, dried and polished as the finish of the pipe or support dictates.

Stainless steel sinks

Stainless steel sinks are to be scrubbed, dried and buffed daily and as necessary throughout the day to ensure that they are kept free from stains

* Courtesy of tenet educational.

and marks. The contractor shall ensure that waste gratings are kept free from obstructions. The external surfaces of taps are to be cleaned with descalar, dried and polished as the tap dictates.

Urinals

All surfaces are to be cleaned daily and as necessary throughout the day with a solution recommended by the manufacturer or other such solution fit for purpose. Particular attention should be paid to flushing cisterns, flush pipes, purge pipes, urinal channels, water management devices and other fittings. To avoid the build-up of uric acid, smears and other stains, the contractor shall employ a programme of regular descaling. The contractor shall ensure that in areas specified by the organisation where there is no automatic disinfectant unit, channel blocks are provided and replenished according to good practice, and sufficiently to ensure that the function of such blocks is achieved. Note: the organisation may have a number of waterless urinals which are required to be cleaned as per the manufacturer's instructions.

Water closets

The internal and external surfaces of pans and associated cisterns and lids shall be scrubbed and dried daily and as necessary throughout the day to ensure that they are kept free from stains and marks, with particular attention being paid to traps. All WC seats and covers shall be disinfected daily, scrubbed, dried and polished. All other WC fittings shall be cleaned as appropriate to the fitting. To avoid the build-up of staining the contractor shall employ a programme of regular descaling. All high level surfaces, door faces, partitions, screens, incidental appliances, fittings and glazing shall be cleaned, dried and polished as the material being cleaned dictates.

Mirrors

Mirrors within WC areas shall be wiped and polished daily and as necessary throughout the day to ensure that the surfaces are free from all smears and marks.

Tiled surfaces

Tiled surfaces shall be washed down, dried and buffed as necessary to ensure that the surfaces are kept free from all stains, drips and marks.

Laminated surfaces

All laminated surfaces are to be wiped, dried and buffed to ensure that the surfaces are kept free of all stains, drips and marks.

Paper towel machines (where applicable)

The contractor is to wipe clean and buff daily and as necessary throughout the day all dispensers where these are installed. The contractor will replenish paper towels.

Electric hand driers (where applicable)

Electric hand driers shall be dry wiped daily and as necessary throughout the day to remove dust, finger marks and drip stains.

Toilet roll dispensers/holders

All toilet roll dispensers and holders are to be wiped, dried and buffed to ensure that they are kept free from stains and marks. The contractor is to replenish toilet rolls as necessary to ensure that a continual provision is maintained.

Toilet brushes and holders

All WCs should be provided with a toilet brush and holder and these shall be kept clean and free of drips and stains.

Waste paper bins/baskets

Waste from all waste paper bins within WC areas is to be removed daily and as necessary throughout the day. The bins are to be damp-wiped to ensure that they are free from stains, marks and drips and the internal surface is also to be cleaned. The waste is to be removed in the bags provided by the organisation and taken to the appropriate receptacle located externally as directed by the FM staff. Waste from such areas is not to be placed into recycling receptacles.

Aids for the disabled

All rails, seats, guards and barriers are to be wiped, dried and buffed to ensure that they are kept free from stains, marks and drips. Specialist WC seats are to be disinfected daily.

Note: all mops, mop buckets, cloths, dusters, leathers, brushes or any other cleaning aid are to be disinfected following their use each day. All such items are to be colour coded or marked to demonstrate that their use is only for within such areas.

Floors/stairs

The contractor is to use their own discretion as to whether mechanical or hand-powered equipment is to be utilised in any instance. Note: during wet-mopping of hard surfaces the contractor is required to position warning signs around the area being mopped; these are to remain *in situ* until the risk of slipping on the wetted surface has passed.

Concrete floors

Concrete floors are to be swept and/or vacuumed as frequently as required to maintain the surface free from stains, marks, dust, debris and chewing gum. Spillages are to be treated with an appropriate cleaning agent, wet-mopped and buffed dry.

Terrazzo floors

To be swept and mopped as frequently as required to maintain the surface free from stains, marks, dust, debris and chewing gum. Spillages are to be treated with an appropriate cleaning agent, wet-mopped and buffed dry. The surface is to be stripped and repolished at intervals recommended by the manufacturer or in accordance with good cleaning practice. Chewing gum is to be removed on a daily basis with an appropriate product, which should not be detrimental to the floor covering.

Vinyl floor coverings

Vinyl floors are to be swept or vacuumed as frequently as is required to maintain a surface free from stains, marks, dust, debris and chewing gum. Spillages are to be treated with an appropriate cleaning agent and wet-mopped. The surface is to be stripped and repolished at intervals recommended by the manufacturer or in accordance with good cleaning practice. Chewing gum is to be removed on a daily basis with an appropriate product, which should not be detrimental to the floor covering. The contractor is to provide the FM Department with risk assessments and method statements for work in such areas.

Floors to workshops and ceramic areas

Floors within such areas should be damp-mopped to minimise the agitation of fine dust. The individual floor covering shall be treated as specified for the particular covering. Chewing gum is to be removed, on a daily basis, with an appropriate product, which should not be detrimental to the floor covering. The contractor is to provide the FM Department with risk assessments and method statements for work in such areas.

Wooden floors

Wooden floors shall be damp-mopped to minimise the agitation of fine dust. The floor is then to be buffed dry. Chewing gum is to be removed on a daily basis with an appropriate product, which should not detrimental to the floor finish. The contractor is to periodically polish and buff wooden floors in accordance with the manufacturer's specifications. The contractor is to provide the FM Department with risk assessments and method statements for work in such areas.

Carpeted areas

To be vacuumed as frequently as required to maintain the surface free from stains, marks, dust, debris and chewing gum. Spillages, stains and chewing gum are to be removed with an appropriate cleaning agent, tested in the first instance in an inconspicuous area.

Walls

Tiled surfaces

All tiled surfaces including grouting are to be washed down as frequently as required to ensure that the surface is free from stains, marks, drips and mould growth. Grouting is to be treated with a fungal inhibitor where fungal growth is encountered.

Wood panelling

To be damp-wiped and polished with an appropriate wood polish as frequently as required to maintain the surface free from stains, marks and drips. Stains and marks should be treated with an appropriate cleaning agent.

Painted wall surfaces

All stains and marks, including graffiti, to be removed using agents, materials and methods that will not damage the wall covering. Such walls should be cleaned as frequently as necessary to maintain the surface free from stains, marks and graffiti. Graffiti is to be removed within 24 hours of notification.

Papered wall coverings

These surfaces should be cleaned according to the demands of the individual surface. In all cases where agents are used, a small area should be tested in an inconspicuous area first. Such walls should be cleaned as frequently as necessary to maintain the surface free from stains, marks and graffiti.

Specialist coatings

These surfaces are to be cleaned as frequently as required to maintain the surface free from stains, marks, grease dust and debris. Such areas are to be cleaned using only agents recommended by the coating manufacturer. The contractor shall instigate a polishing programme in accordance with the manufacturer's specific instructions and at intervals to suit the covering.

Internal surfaces of lift cars and external door faces

To be cleaned with an appropriate stainless steel cleaning agent and polished to a high lustre. Panels, switches and rails are to be damp-wiped, dried and polished with an appropriate agent at a frequency required to ensure all surfaces are maintained free from all stains, marks and smears and which complements the natural lustre of the surface. Floors are to be cleared and graffiti and chewing gum removed daily.

Ceilings

Other than by specific request and under separate contract and cost, the organisation shall not require any ceilings to be cleaned.

Doors

Internal doors (all types)

All doors are to be damp-wiped, dried and dry-buffed to ensure that they are free from all stains, marks, smears, dust and dirt. Periodically, doors are to be polished on both sides to maintain the natural lustre of the surface. Polish is to be of a make suitable for the individual surface. The frequency required is that needed to ensure that doors are maintained free from all stains, marks and smears.

External doors (all types)

As above, but thresholds are to be brushed out and damp-wiped.

Door furniture

All door furniture, including closers, handles, latches. openings, fire seals, signs, locks, kick-plates, push-plates, hinges, floor springs and stops, are to be damp-wiped, dried and buffed to remove all stains, marks, dust and debris. Door furniture is to be periodically polished to maintain the natural lustre of the individual surface. Polish is to be of a make suitable for the individual surface. The frequency required is that needed to ensure that door furniture is maintained free from all stains, marks and smears.

Door glazing (all types)

All door glazing, irrespective of material, is to be damp-wiped, dried and dry-buffed to ensure that the surface is free from all smears, dust and marks. Periodically, the contractor shall apply an agent which polishes and leaves a smear-free finish. Polish is to be of a make suitable for the individual surface. The frequency required is that needed to ensure that door glazing is maintained free from all stains, marks and smears.

Internal partitions and screens

Internal screens

All tops, sides, ledges and glass components are to be damp-wiped, dried and dusted to remove all stains, marks and smears. The contractor is to ensure that glass components are cleaned with an appropriate agent which shall ensure a smear-free finish to both sides. The frequency required is that needed to ensure that screens and partitions are maintained free from all stains, marks and smears.

Internal partitions (structural)

The surface is to be cleaned in a manner to suit the finish.

Internal partitions (fabric-covered)

To be brushed and or vacuumed to remove dust, dirt and debris. Stains should be damp-wiped with an appropriate agent recommended by the manufacturer. The frequency required is that needed to ensure that partitions are maintained free from all stains, marks, dust and debris.

Furniture

Chairs: vinyl/polypropylene

All chairs are to be damp-wiped and dry-polished as frequently as required to ensure that the surfaces are kept free from stains, marks and dust.

Chairs: upholstered

All upholstered chairs are to be vacuumed and/or brushed to maintain the surface free from dust, debris and marks. Any stains should be treated with an appropriate agent recommended by the manufacturer, with a sample test being carried out in an inconspicuous area first. Chair legs and frames are to be damp-wiped and dry-polished to maintain the surface free from stains, marks and dust.

Metal furniture

Frames and surfaces are to be damp-wiped and polished to ensure that the surfaces of all components are kept free from stains, marks and dust. Cleared surfaces only are to be cleaned; the contractor is not to move items or equipment to clean.

Wooden furniture: desks, pedestals, drawer units, coffee tables, cabinets, cases

All wooden furniture is to be cleaned with the cloth and polish recommended by the manufacturer. The frequency should be such to ensure that the surfaces are kept free from stains, marks, dust and debris. Cleared surfaces only are to be cleaned; the contractor is not to move items or equipment to clean.

Workbenches/worktops

All workbenches/worktops are to be cleaned and polished in a manner which reflects their use.

Glass surfaces

All glass surfaces are to be cleaned with an appropriate agent that will not leave stains, smears, smudges or dust. The frequency of cleaning shall be as necessary to maintain the surfaces free from smears, smudges or dust. Cleared surfaces only are to be cleaned; the contractor is not to move items or equipment to clean.

Audio-visual, telecommunications and other specialist equipment

The contractor is not required to clean any of these items. The organisation's unit technicians maintain such equipment. Such equipment should not be moved, tampered with or disconnected by the contractor.

Computer equipment

the contractor is required to light dust the computer base units only where accessible. More intensive cleaning of computer equipment is not included as part of the cleaning specification.

Gym equipment

The contractor is only to clean equipment that is electrically isolated and switched off at the mains. All gym equipment shall be damp-dusted to remove all marks, stains and drips. The frequency should be as necessary to maintain the equipment free from all stains, marks and drips.

Mirrors in all areas and locations

All mirrors, irrespective of material, are to be damp-wiped, dried and dry-buffed to ensure that the surface is free from all smears, dust and marks. The contractor shall periodically apply an agent which polishes and leaves a smear-free finish. Polish is to be of a make suitable for the individual surface. The frequency required is that to ensure that all mirrors are maintained free from all stains, marks and smears.

Curtains and blinds

Window curtains and blinds

Curtains and blinds (horizontal and vertical) are to be vacuumed and/or dusted as frequently as required to maintain the material free from dust, dirt, marks and debris. The contractor is not required to remove blinds or curtains from fixings for the purpose of cleaning or clean above 1.8 m high (other than for shower curtains).

Shower curtains

Shower curtains are to be wiped down with a damp cloth to ensure that all stains, marks and drips are removed. The frequency should be as necessary to ensure that the curtains are maintained free from all stains, marks and drips. Shower curtains are to be periodically unclipped from hangers and washed in a solution recommended by the manufacturer, dried and re-hung on existing hangers.

Skirting, ledges, shelves, reveals, architraves and mouldings, sills, trunking and switches up to 1.8 m high

All areas and locations are to be damp-wiped and/or dusted as required to maintain surfaces free of dust, stains, marks and smears. The contractor shall periodically apply an agent or polish which leaves a smear-free finish. Polish is to be of a make suitable for the individual surface. The frequency required is that to ensure all surfaces are maintained free from all stains, marks and smears and complements the natural lustre of the surface.

Radiators, convectors, pipes, valves, fire appliances and hand rails

All areas and locations to be damp-wiped and/or dusted as required to maintain surfaces free from dust, stains, marks and smears. The contractor shall periodically apply an agent or polish which leaves a smear-free finish. Polish is to be of a make suitable for the individual surface. The frequency required is that to ensure that all surfaces are maintained free from all stains, marks and smears and complements the natural lustre of the surface.

Waste bins

Black and white bin liners

Black and white bin liners are to be removed from the relevant building and placed within the general waste container located externally. Black and white bin liners must not be placed within the recycling containers. The waste bins are to be emptied daily, damp-wiped as required to ensure that the bins are free from stains, marks and dirt.

Cardboard boxes and packaging

These should be removed as necessary throughout the day. Boxes must be flattened before placing in the container provided.

Stainless steel hand rails, brackets, feet, fittings and bins

These are to be cleaned with an appropriate stainless steel cleaning agent and polished to a high lustre. The frequency required is that to ensure all surfaces are maintained free from all stains, marks and smears and complements the natural lustre of the surface.

Bins (all levels)

The bins are to be emptied daily and as necessary throughout the day. The bins are to be damp-wiped, dried and dry-buffed to ensure that the surface is free from all smears, dust and marks. The contractor shall periodically apply a polish which leaves a smear-free finish. The polish is to be of a make suitable for the individual surface.

Internal exposed steel frame members up to 1.8 m high

The frame members should be damp-wiped, dried and buffed to remove all dust, debris, dirt, stains and marks. The frequency required is that to ensure that the frame members are maintained free from all stains, marks and smears. The contractor should provide risk assessments and method statements to the FM Department for all work connected with cleaning this element.

Schedule 3: Performance standards

Table 1. This table summarises the standard of cleaning expected by the organisation

Item	Required standard	Unacceptable standard
Balustrades, handrails	Surfaces shining and free from all dust, grime, grease, fingermarks and other scuffs and marks	Build-up of dust and grease, scuffs, fingermarks or grime on horizontal surfaces or in corners
Bath and basin supports	Removal of all dust and dirt	Ingrained dirt, deep covering of dust

(Continued)

Table 1. (*Continued*)

Item	Required standard	Unacceptable standard
Blinds: vertical, horizontal or roller	Free from dust and dirt	Heavy build-up of dust and dirt
Bookcases and display cabinets	Removal of all dirt and dust, surfaces left in clean and shiny condition as appropriate; glass shiny and smear-free	Build-up of dirt and grease, heavy fingermarks, dulled wood finish
Cabinets, filing cabinets, lockers, cupboards	Removal of all dirt and dust, grease and fingermarks; clean handles	Build-up of dirt and grease, heavy fingermarks, soiling, powder
Carpets, rugs and foot mats	Free from daily debris, stains and dust; spot removal carried out if necessary; fresh smell	Soil build-up within pile or accumulation of fluff and other dirt/dust, especially under desks, around the edges of furniture, in corners and other areas of difficult access; dried spots and stains which are removable; matted pile
Ceramic tiles	Clean, grease- and smear-free and buffed shiny	Smears, ingrained dirt, black-encrusted grouting, build-up of soiling
Chairs	All soiling and dust removed	Sticky to touch, build-up of dust and dirt on frames, chewing gum, discoloration or darkening of seat, back and arms
Charts, wall pictures	Free from dust, grease and fingermarks	Build-up of fingermarks, dust smears on glass
China, glass and cutlery	All food particles removed and in a dry, non-greasy condition; glass sparkling clean	Food particles, grease or other dirt adhering to article, smears on glass or discoloration/build-up of algae
Dispensers, incinerators and disposal units for sanitary items	Removal of soiling and ash; surfaces wiped clean, dry and bright	Build-up of dirt and dust, ash deposits, grease, etc.; discoloration of outer surfaces, stains and excessive fingermarks

Item	Required standard	Unacceptable standard
Doors, including kick- and hand-plates	Clean, shiny and dust-free; no fingermarks	Build-up of dirt, dust, grease or fingermarks; smears on glass, heavy scuffing to base or kick-plate
Draining boards	Removal of all food particles; dry and polished	Heavy water stains, soiled or stained surface and corners
Drinking fountains	Removal of soiling, limescale or smears from inner and outer surfaces; dry and polished	Smears and stains on inner or outer surfaces, dirt in plugholes, smears, limescale, discoloration
Fire appliances	Dust-free	Visible dust, smears, stains and fingermarks
Floors, polished hard floors, including lifts	Removal of soil and debris arising from daily activities; removal of spots, soil build-up, scuff marks; clean, dry, smear-free and shiny	Litter, dust, fluff, dirt and other debris on the surface with accumulation in the corners; traffic marks, mop marks, streaks and removable stains; ingrained/impacted dirt, particularly in crevices, corners and around fittings; detergent residue; build-up of dirt and dust in the guides and runners of lift doors
Flush, pipes and cisterns	Removal of all soiling and stains leaving the surfaces and handle hygienically clean	Dust, grime, rust, smears or run marks
Furniture, including noticeboards, fire appliances, plant holders	Removal of all dust, dirt, debris, fluff; surfaces cleaned and treated according to type of finish; no fingermarks, smears, grease marks	Build-up of dust and ingrained dirt, particularly in corners; grease, smears heavy fingermarks, fluff on flock topped tables; dulled wooden surfaces; polish and watermarks around the base from floor cleaning; graffiti
Glass and glazed surfaces, including mirrors	Removal of all dirt, dust, debris; polished clean and shining without smears or streaks	Heavy build-up of dust, fingermarks, smears, particularly in corners; overall film of grease

(*Continued*)

Table 1. (*Continued*)

Item	Required standard	Unacceptable standard
Lampshades (up to 1.8 m above floor level)	Removal of all dust and other types of soiling; polished fittings	Heavy build-up of dust and fingermarks; insect marking, smears and stains, tarnished fittings
Ledges, shelves, skirting	Removal of all dust and soiling, surfaces clean and, if applicable, shiny	Heavy build-up of dust, smears and stains
Mats at entrances	Clean, free from dust, dirt and all debris; fresh smell	Build-up of dirt around edges, underneath, or in the well; clogged with dried mud or other debris
Overflows	Removal of all visible stains, soiling and limescale build-up	Soiling visible on the inside of the overflow, build-up of limescale, discoloration by ingrained dirt or dust
Paintwork (usually gloss)	Removal of all types of dirt, dust and stains leaving the surfaces clean and shining	Build-up of stains and dirt, especially in the corners; film of grease
Partitions and doors	All surfaces wiped dry, clean and bright; no smears, stains, fingermarks or graffiti	Dust, dirt, stains, heavy fingermarks, graffiti
Pipes and surrounds	Clean, dry and free from dust and dirt	Dusty, stained and soiled
Pipes, radiators, storage heaters, towel rails	Removal of all dirt, dust and stains leaving the surfaces clean, bright and shiny	Build-up of dirt, dust or stains, marks made as a result of cleaning floors, walls, grease or fingermarks
Plugs and chain	Removal of all dust and dirt	Film of grease, build-up of soap or limescale
Roller and paper towel cabinets	All surfaces clean and bright, wiped free of fingermarks and other stains, sufficient supply of towels to last until next refill	Build-up of dirt, dust, fingermarks and other stains; no towels

Item	Required standard	Unacceptable standard
Shower curtains	Removal of all stains and watermarks; curtains may require occasional washing	Staining from old soap, dust, talcum powder
Shower gullies	Removal of all soiling leaving surfaces hygienically clean	Accumulated grease and soiling, hair in drain holes
Showers, hand basins, sinks	Ceramic and steel left in a dry and shiny condition with watermarks removed from edges	Tidemarks, scum, limescale, build-up of soap deposits, removable discoloration
Soap dispensers	Surfaces free from dirt, dust, soap deposits and wiped dry and bright	Dust and dried soap adhering to handle; dulled surface if glass or chrome
Taps	Removal of all dirty marks and limescale, polished with dry cloth to enhance finish	Chrome finish taps wholly or predominately dull (having regard to age), soap or limescale encrustation, dirt or grime at base of taps
Telephones	Removal of all soiling, leaving all surfaces including the mouth and earpiece in a hygienically clean and shiny condition	Build-up of dirt, dust and stains; greasiness and other accumulated dirt in mouthpiece and earpiece
Toilet floors	Clean, free from soil, dust, debris and other matter arising from the day's use; dry and, where applicable, shiny	Mop marks, build-up of dirt, dust, soil across floor and around bases of pans, sinks and other fittings; stains
Toilet roll holders	Clean, sufficient toilet tissue available to last until next refill	Surface covered in dust and stains; no toilet tissue
Toilet pans	Removal of all soiling and stains leaving the pan in a hygienic condition and ceramic surfaces bright and shiny; fresh, pleasant smell	Dust, deposits, watermarks, soiling under rim, powder deposits in bend, unpleasant smell

(*Continued*)

Table 1. (*Continued*)

Item	Required standard	Unacceptable standard
Urinals	Removal of all stains, deposits, watermarks from walls, gullies, corners and overhangs; surfaces wiped clean and bright; pleasant smell	Limescale and other deposits, especially under rims in channels and traps; stains on all surfaces; build-up of dust and dirt; unpleasant smell
Wastepaper receptacles, bins, litter bins	Removal of all litter and rubbish, wiped clean inside and out; where applicable clean bag inserted; fresh smell	Deposits of dirt adhering to any surface; smears, stains, unpleasant smell
WC seats and covers	Removal of all dirt, dust and stains; seat left in a hygienic condition	Dirty marks adhering to the top or underside, limescale deposits, build-up of dirt around hinges

Table 2. Required frequency of cleaning

Area or type of area	Frequency
Offices	Cleaned once each week; bins to be emptied on a daily basis
Functional spaces	Cleaned on a daily basis, Monday to Friday
Corridors, foyers and entrances	Cleaned on a daily basis, Monday to Friday
Toilets	Fully cleaned twice a day, with an interim clean/check at one hour intervals
Technical areas/workshops	Cleaned once each week
Catering units (not kitchens)	Cleaned on a daily basis, including tables and floor areas; tables to be moved once each week and full floor buffed/mopped
Offices of Directors/senior managers	Cleaned daily

Note: Special cleaning. A number of areas need an annual strip and seal to maintain the flooring condition.

Appendix II
Grounds maintenance specification for a college

General description

This specification should be read in conjunction with the Preliminary Particulars.

The contractor is requested to provide a fixed-price tender for the landscape maintenance of the College grounds. Namely, the following buildings and surrounding area: Building A, Building B, Building C and Building D.

Specification

1 Grass areas

1.1 The frequency of all grass cutting operations shall be as detailed within these documents for the whole site.

1.2 No growth regulators or herbicides of any form shall be applied to any area of turf at any time unless instructed by the Facilities Manager.

1.3 During periods when ground conditions are too wet to allow grass cutting without causing damage to the surface or levels of the ground, or producing divots, the Contractor shall cease grass cutting operations, notifying the Facilities Manager immediately of this action.

1.4 Mowing shall not be carried out when frost is on the grass.

1.5 Should the Contractor cause damage to the surface or levels of the ground, or create divots during grass cutting operations, the Contractor shall reinstate such damage to the satisfaction of the Authorised Officer.

1.6 Before every cut the Contractor shall inspect all areas to be mowed and shall remove all litter debris and obstructions, e.g. stones, which might cause damage or injury, all such items being collected and removed to the Contractor's tip. This work should be coordinated with any grass cutting activity to ensure that no more than one hour elapses between the litter collection and the commencement of mowing. Under no circumstance will the heights of cut be varied from the Specification to compensate for failure to undertake this task.

1.7 Mowing around obstructions and in the proximity of margins shall be undertaken at the same standard and frequency as that applied to the main area, using methods, tools and machines as appropriate. The cutting of such areas shall be undertaken within 24 hours of the cutting of the main areas.

1.8 Care shall be taken to avoid damage to trees and shrubs. Any damage caused shall be made good as directed by the Facilities Manager to his or her entire satisfaction. Where strimmers are used they shall be fitted with a guard to ensure that no damage is caused to the bark of trees. Under no circumstances will herbicides be used to manage growth around trees.

1.9 Mowing shall be carried out as close as possible to fixed obstructions, moveable obstructions shall be removed to facilitate cutting and replaced before the Contractor leaves the site.

1.10 The Contractor shall at all times during the period of the Contract ensure that all machines engaged in grass cutting operations are sharp and properly set to produce a true and even cut. At no time shall mowers be adjusted so that "scalping" takes place.

1.11 All mowing machines, gang units, etc. are to have grass cuttings removed before proceeding onto roads or paths within the College site.

1.12 The Contractor shall at all times ensure that machines are properly guarded in accordance with the manufacturer's stipulations and maintained to present no danger to the operator or any person in the vicinity of operations.

1.13 It is essential that machinery of all kinds is used in such a manner as not to cause apprehension, damage to a person's belongings or injury to persons. The Contractor is reminded that on the site, staff and visitors have the right of way.

1.14 On occasion, maintenance schedules will be curtailed by examinations or other activities where noise would be disruptive. When this takes place, the Contractor is to arrange for any necessary grass cutting to be completed as soon as notification is received that work may recommence in the area before resuming his normal mowing schedules. On no account shall such areas be left until the next cutting or maintenance cycle.

1.15 In certain areas flowering bulbs have been planted or have naturalised. In such areas mowing shall cease prior to the emergence of bulbs and the Contractor shall not cut the grass until a minimum period of eight weeks has elapsed after completion of flowering, this being agreed prior to cutting with the Facilities Manager.

1.16 Where in the view of the Facilities Manager work has been carried out to an insufficient standard, the work shall be carried out again by the Contractor without creating a backlog to other maintenance schedules.

1.17 In normal weather conditions the Contractor will be required to keep to the planned programme. Variations to the grass cutting schedules may arise if weather conditions create additional growth to the extent that additional cuts are required. In this case the Facilities Manager will issue instructions for extra work and the Contractor will need to vary the planned programme.

1.18 If inclement weather prevents work being carried out the Contractor will be required to notify the Facilities Manager. The Contractor will be required to resume work as soon as possible in accordance with the programme. Replacement work will need to be found and agreed with the Facilities Manager.

1.19 If weather conditions inhibit the growth of grass to negligible amounts between scheduled cuts the Facilities Manager may prolong the interval between cuts or suspend all cutting operations. Replacement work will need to be agreed with the Facilities Manager.

1.20 Hand-held mechanical cutters may be used where grass or meadow areas are inaccessible to vehicles or wheeled mowers or are too steep, too small, too uneven or obstructed to be cut using other machines.

1.21 The slopes of banks and ditches may be cut using a Flymo or a strimmer.

1.22 Prior to the commencement of the contract the Contractor shall provide if requested full details of all risk assessments in relation to grass cutting including all method statements relating to all operations.

2 Cylinder/rotary mowing, no grass collection

2.1 The area shall be cut leaving no areas uncut between rows and producing an even height across the whole area.

2.2 The period of time between cuts shall unless otherwise instructed or delayed by ground conditions be no greater or less than the instructions within the documents.

2.3 Grass clippings shall lie where they fall except in the case of those falling on hard surfaces. In such cases these shall be swept or removed by air blower from the surface before the Contractor leaves the site.

3 Cutting variations. On occasions the Contractor will be required to vary the cutting height, carry out additional cuts, or complete the work earlier or later for special events at the College.

4 Grass cutting heights and frequencies of cuts. The locations for the different cutting regimes are shown on the provided plans.

4.1 Grass Cut 1. Cut grass with cylinder mower to 25 mm at weekly intervals between mid-March and the end of October. To include all follow-up mowing. Note: the follow-up mowing to include all banks, obstacles and any rough grass areas beneath trees within the area designated for weekly cutting. Total number of cuts = 33.

4.2 Grass Cut 2. Cut grass every two weeks by cylinder or rotary boxed mower to 25 mm. Total number of cuts = 16.

4.3 The Contractor should note that the measured areas containing bulbs will not require cutting until instructed by the Facilities Manager. These areas will then return to scheduled cuts.

5 Grass edging

5.1 Annual maintenance

5.1.1 Once a year during winter the Contractor will reform with a half-moon edging tool or approved mechanical edger the edge of grass areas where it abuts footpaths, roads and other areas within the Inner Site.

5.1.2 Care should be exercised that only the minimum amount of turf is removed conducive with the forming of a true edge. All waste turf and soil to be swept up and removed to Contractor's tip at the end of each working day.

5.2 Routine maintenance

5.2.1 Between mid-March and October inclusive grass edges around beds and open borders within the Inner Site shall be cut with long-handled shears or a suitable mechanical implement at three week intervals.

5.2.2 All waste from grass edging on Inner Site to be removed to Contractor's tip at the end of each working day.

6 Shrub beds

6.1 General. Beds throughout the site are planted with a mixture of deciduous and evergreen shrubs. Those containing shrub and species roses will be treated as shrub beds.

6.2 Winter maintenance. In February the Contractor shall undertake the following operations.

6.2.1 Cut back shrubs to prevent encroachment onto footpaths or roads and reduce in height not to exceed 1 m to prevent the blocking of light from windows. All waste to be removed to Contractor's tip at the end of each working day.

6.2.2 Remove all litter and debris.

6.2.3 Lightly fork soil to a depth of 75 mm removing weeds including the roots of perennial weeds.

6.2.4 Apply a mulch of well-composted material or bark (not wood chippings).

6.3 Routine maintenance. From April to October shrub beds will be visited at four-week intervals and the following work shall be undertaken.

6.3.1 Remove weeds including roots (no strimming).

6.3.2 Carry out encroachment pruning.

7 Hedges

7.1 General. Hedges are to be cut using hand shears or mechanical machinery. Full health and safety risk assessments must be undertaken on the use of all machinery on site. A full copy of the risk assessment and method statements must be made available to the contract administrator before the start of the contract.

7.2 Cutting. In general, all surfaces shall be cut back to the previous year's growth in June and September, ensuring no encroachment to paths and roads takes place. This activity shall be carried out by a skilled operative and the Contractor must ensure that all shears, hand or mechanical, are sharp. The finished cut face shall be "square" with a slight taper from the bottom of the hedge to the top. Any clippings lodging in the top or sides of ornamental hedges shall be removed by hand picking as required.

7.3 Hedge bases

7.3.1 Winter maintenance. In February all hedge bases will be cleared of weeds, brambles and debris and a mulch of well-composted material or bark (not wood chippings) applied.

7.3.2 Routine maintenance. Between April and October at four-week intervals all hedge bases will be cleared of any weed re-growth.

8 Plant containers

8.1 External plant containers. Planting containers of annuals: twice each year in autumn and spring, and when instructed, plant out external plant containers around any perennial plants.

8.1.1 Clear the previous season's bedding, including all bulbs and remove all waste.

8.1.2 Dig over the containers, removing all trash and annual weeds. Carefully dig out perennial weeds and remove waste. Either in autumn or spring (as instructed) remove haulm and incorporate compost or fertiliser provided by the Contractor.

8.1.3 Plant out annual bedding plants, corms, bulbs or tubers in accordance with good horticultural practice. Keep plants watered as necessary before planting out and water in after planting.

8.1.4 Approval must be obtained for any replanting which may be required, the plants being provided by the Contractor. Should the loss of plants be due to the Contractor's negligence in horticultural practice or otherwise, the cost of the replacement plants will be met by the Contractor.

9 Car park maintenance

9.1 General. In general, the maintenance work specified for individual operations will apply at the frequencies designated for the area in which the car park is situated (litter scavenge, leaf collection, etc.).

9.2 Work will need to be coordinated with the usage of the parking facilities and it may be necessary for the work to be undertaken during hours that the car parks are not fully in use. However, all will be undertaken within normal working hours.

9.3 Winter maintenance. During the winter the perimeter of car parks will be maintained as follows.

9.3.1 All litter debris and rubbish will be removed to the Contractor's tip.

9.3.2 Brambles and other invasive undergrowth will be cut back to ground level and treated with an approved herbicide.

9.3.3 Any encroaching shrubs, trees or hedges will be cut back to allow the full extent of the parking bays and thoroughfares to be utilised.

9.3.4 The use of tractor-mounted machinery is not permissible.

10 Leaf collection

10.1 From the second week in October until mid-December leaves will be cleared by the Contractor from within the whole of the area of the Inner Site and from paths and steps around buildings including the residences and taken to the Contractor's tip.

10.2 From the first week in November the whole site will be cleared of leaves ensuring that any accumulation of leaves on the edge of woods and copses are removed so as not to spread to areas already cleared.

10.3 The frequency of visits shall be at the discretion of the Contractor, but if sites are not substantially clear of leaves at the end of a payment period the Contractor shall not be paid.

10.4 All sites to be cleared of leaves by the end of December.

10.5 The Facilities Manager will provide a site for the disposal of cleared leaves suitable for composting to allow the Contractor to recycle the end product.

12 Site scavenge

12.1 The scavenge is to consist of the removal to the Contractor's tip of all extraneous rubbish not arising from the day-to-day workings, including leaves, stones, bricks, debris, paper, bottles, cans, tipped material, dead wood, branches and other tree debris and any other materials whatsoever their composition and by whomsoever deposited that the Facilities Manager considers detrimental to the appearance of the site.

12.2 The Inner Site will be scavenged daily on each working day. On some occasions such as Open Days it may be necessary for extra clearances. At each scavenge litter bins will be emptied.

12.3 The remaining site will be scavenged weekly. Litter bins will be emptied three times each week.

12.4 Skips for the disposal of waste from the litter scavenge and emptying of litter bins will be provided by the Contractor.

13 Maintenance of roads, paths and hard surfaces

13.1 General. The Contractor will be responsible for the following work on all roads, paths and hard surfaces within the Site.

13.1.2 Sweeping and clearing of rubbish and debris.

13.1.3 Keeping the surfaces free of weeds.

13.1.4 Paths and hard surfaces will be swept weekly by mechanical means or, in areas that are inaccessible to larger machinery, e.g. steps, by hand-held broom or air blower.

13.1.5 Roads will be swept weekly along both kerb lines by mechanical means.

13.2 Weed control. The Contractor will keep all hard surfaces free of weeds throughout the year to the reasonable satisfaction of the Authorised Officer.

13.2.1 Weeds can be spot treated with an approved translocated herbicide applied either by knapsack sprayer or CDA applicator. All risk assessments in relation to the use of herbicides shall be undertaken in advance of the commencement of the contract and made available to the Contract Administrator.

13.2.2 Where hard surfaces abut grass areas any grass encroaching onto the hard surface will be treated.

13.2.3 All dead grass and weeds will be removed to the Contractor's tip. Under no circumstances shall the Contractor use the organisation's waste receptacles on site.

14 Small tree maintenance. General. The work detailed below is applicable to trees not exceeding a girth of 350 mm measured 1 m from the ground. Trees with a girth in excess of 350 mm shall be treated under arboriculture contracts.

14.1 Winter maintenance shall take place in February and will comprise the following work.

14.1.1 Soil bases of trees in grass will be kept free of grass and weeds in a circle of 600 mm diameter by a mulch of well-composted material or bark (not wood chippings).

14.1.2 Crowns will be lifted to provide headroom of 3 m. Branches are to be cut back 500 mm from the kerb lines adjacent to roadways; all overhanging branches that could get caught by visitors, staff or vehicles to be removed/ cut back to a safe height. Crown prune by removing dead branches by one-third to present a well-balanced head, in each case cutting back to live wood.

14.1.3 Stakes and ties shall be checked and adjusted, replaced or removed as required.

14.2 Summer maintenance. In June and August the Contractor shall undertake the following tasks.

14.2.1 Remove basal and epicormic growth with suitably sharpened hand tools.

14.2.2 Stakes and ties shall be checked and adjusted or removed as necessary.

15 Grass seeding

15.1 Where grass seed is to be sown the Contractor will fork over the area to a depth of 100 mm or cultivate to depth using an approved machine as appropriate to the size of the area. The soil is then to be trodden firm, raked level and cultivated to a fine tilth. All stones exceeding 25 mm gauge, all weeds and other debris shall be removed from site. Level by Trulute and make up hollows with soil as specified by the Facilities Manager.

15.2 Where finished levels are not given, the levels shall be such that the finished surface will be smooth with an even fall between the boundaries of the area and free of hollows in which ponding can occur.

15.3 The Contractor will apply a pre-seeding fertiliser, raked in three to seven days before the commencement of seeding. All seed will be as specified and will be supplied by the Contractor and applied at the instructed rate. The seed will be sown in two equal applications over the areas in transverse directions to ensure an even distribution, seed is then to be raked or harrowed in evenly over the area. The seeding of areas will only take place as directed.

16 Turfing

16.1 Where a portion of existing grass is to be re-turfed, the entire area shall be cultivated and thoroughly broken by mechanical or manual means to a uniform depth of 100 mm. All stones exceeding 25 mm gauge, all weeds and other debris shall be removed from site. Excessive soil shall be removed from site to achieve a compacted level of 40 mm below the adjacent grass. The bed shall be compacted to true level by treading and raking.

16.2 All turf will be as specified and supplied by the Contractor. Unloaded stacks of turf shall not exceed 1 m in height, stacked soil-to-soil, and shall not stand longer than five days.

16.3 The turfs shall be laid so as to break bond by half a turf length, be neatly cut to edges and around obstructions, tightly butted to each other and the surrounding grass and well firmed down.

16.4 Deviations from the specified levels shall be adjusted by making up levels with fine soil as specified, or by lifting and raking out surplus soil. Turf shall not be firmed by use of beaters or heavy rollers.

16.5 Immediately after laying turf the Contractor shall top-dress with soil as specified and brush this well into the joints. The Contractor will be responsible for ensuring the turfed area receives sufficient water for the grass to become established. Turf shall not be laid in very dry or frosty weather.

16.6 Turf on banks and steep gradients shall be secured with U-shaped, 200 × 4 mm diameter galvanised wire pins or wooden pegs driven through the turfs into the bed.

Appendix 12
Health, safety and environment questionnaire

Section A: Health and safety

Details of Contractor

Name of Company:

Address:

Tel No.:

Fax No.:

Email address:

Declaration

I certify that the details given in this questionnaire are correct and accurate.

Name (block capitals):

Job title:

Signature:

Date:

Nature of business

Numbers of staff employed:

Indicate the type of work/services you provide and for which contact/project you wish to be considered:

Health and safety policy

Do you have a health and safety policy?

Yes/No/NA (if yes, please attach a copy):

Who is the person with overall responsibility for health and safety matters within your company?

Name:

Job title:

Contact details:

Health and safety surveillance and advice

Who is the person(s) employed to ensure your company meets the requirements of health and safety legislation and that your employees adopt safe working practices (e.g. H&S Manager/Advisor/Consultant)?

Name:

Job title:

Qualifications:

Length of experience in role:

Contact details:

Health and safety training

Are your employees and subcontractors trained in their legal duties, your company rules and your safe working procedures?

Yes/No:

If the answer to the above is "yes", are the training records available for inspection should this be necessary?

Yes/No:

Accident investigation records

Are all accidents/incidents involving your employees investigated?

Yes/No:

If yes, by whom:

Accident statistics (previous two full years)

No. of reportable accidents involving your employees:

Last year:

Previous year:

Total no. of accidents involving your employees (reportable and non-reportable):

Last year:

Previous year:

Safe systems of work

Please give brief details of appropriate safe systems of work which you have developed and which are applicable to the work you wish to undertake at this Company (e.g. Permits to Work, Method Statements, Health and Safety Manual, Safe Working Procedures, Risk Assessments):

Are these systems available for inspection should this be necessary?

Yes/No:

Plant, equipment and vehicles

How do you ensure that that plant, equipment and vehicles to be used on site are in good condition and safe to use?

Do all drivers of vehicles hold current and valid licences?

Yes/No:

Are these licences available for inspection should this be necessary?

Yes/No:

Insurance

Do you have insurance covering the following areas?

Employers' liability insurance?

Yes/No/NA:

Name of insurers:

Public liability insurance?

Yes/No/NA:

Name of insurers:

Subcontractors

How do you assess the health and safety record and competence of companies with whom you place contracts and who may be used by your company on our site?

Have you verified that the subcontractors hold the appropriate insurance cover?

Yes/No:

Enforcement action

Please details any prosecutions or improvement/prohibition notices that have been served on your Company within the last five years:

Health and safety awards

Please give details, with dates, of any health and safety awards received:

Section B: Environment

Environmental management

Does your company operate a certified Environmental Management System (EMS) (e.g. ISO 14001, EMAS)?

Yes/No:

If yes, please detail awarding body, dates, etc.:

If your company operates a current certified EMS, this questionnaire is now complete.

If your company does not yet operate a current certified EMS, please complete the remaining questions.

Environmental surveillance and advice

Who is the person(s) employed to ensure your company meets the requirements of environmental legislation and that employees adopt environmentally sound practices (e.g. Environmental Manager/Coordinator)?

Name:

Job title:

Qualifications:

Length of experience in role:

Contact details:

Hazardous and environmentally harmful substances

Do you monitor and record the usage of hazardous or environmentally harmful substances?

Yes/No:

If yes, please give details:

Waste disposal

Do you have records of wastes you have disposed of?

Yes/No:

Are these records available for inspection should this be necessary?

Yes/No:

Do you have a waste disposal procedure?

Yes/No:

Are your employees aware of this procedure?

Yes/No:

Energy and waste minimisation

Do you have a waste and energy minimisation programme?

Yes/No:

If yes, please give details:

Does your organisation have a carbon plan?

Yes/No:

If yes, what are your targets for the coming year?

Environmental enforcement action

Please detail any prosecutions, notices or warnings received within the last five years:

Environmental awards

Please give details, with dates, of any environmental awards received:

Thank you for taking the time to complete this questionnaire

Please return the completed questionnaire to:

Name:

Job title:

Copies of this questionnaire will not be returned to you, nor passed to non-interested parties.

References

Buchanan, D.A. and Huczynski, A.A. (2010) *Organizational Behaviour,* 7th edn. Pearson Education, Harlow.

CIBSE *Heating, Ventilating, Air Conditioning and Refrigeration* (CIBSE Guide B). Product Code GVB.

HSE (2012) *Getting to Grips with Manual Handling. A Brief Guide.* Leaflet Series INDG 143 (REV 2), HSE Books.

HSE (1997) *Electrical Safety at Places of Entertainment.* Series code GGS50, HSE Books.

HSE (1998) *Lighting at Work,* 2nd edn. Health and Safety Guidance Series HSG38, HSE Books.

HSE (1999) *Simple Guide to the Lifting Operations and Lifting Equipment Regulations 1998.* Leaflet Series INDG 290t (web only), HSE Books.

HSE (2000a) *Legionnaires' Disease. The Control of Legionella Bacteria in Water Systems. Approved Code of Practice and Guidance,* 3rd edn. Legal Series L8, HSE Books.

HSE (2000b) *Safety in Working with Lift Trucks,* 3rd edn. Health and Safety Guidance Series HSG6, HSE Books.

HSE (2002) *Upper Limb Disorders in the Workplace,* 2nd edn. Health and Safety Guidance Series HSG60, HSE Books.

HSE (2003a) *Electricity at Work. Safe Working Practices,* 2nd edn. Health and Safety Guidance Series, HSE Books.

HSE (2003b) *Manual Handling Assessment Charts.* Leaflet Series INDG383, HSE Books.

HSE (2005a) *Controlling Noise at Work. The Control of Noise at Work Regulations 2005. Guidance on Regulations.* Legal Series L108, HSE Books.

HSE (2005b) *Noise at Work. Guidance for Employers on the Control of Noise at Work Regulations 2005.* Leaflet Series INDG362 (REV1), HSE Books.

HSE (2005c) *Preventing Slips and Trips at Work.* Leaflet Series INDG225 (REV1), HSE Books.

HSE (2005d) *Workplace Transport Safety: An Employers Guide.* Health and Safety Guidance Series HSG136, HSE Books.

HSE (2005e) *Workplace Transport Safety. An Overview.* Leaflet Series INDG 199 (REV1), HSE Books.

HSE (2006a) *Asbestos. The Licensed Contractor's Guide.* Health and Safety Guidance Series HSG 247, HSE Books.

HSE (2006b) *Work with Materials Containing Asbestos. Control of Asbestos Regulations 2006. Approved Code of Practice and Guidance.* Legal Series L143, HSE Books.

HSE (2007a) *COSHH Essentials: Easy Steps to Control Chemicals. Control of Substances Hazardous to Health Regulations.* Health and Safety Guidance Series HSG 193 (web only), HSE Books.

HSE (2007b) *The Work at Height Regulations (2005) (as amended) A Brief Guide.* Leaflet Series INDG401, HSE Books.

HSE (2008a) *Clearing the Air. A Simple Guide to Buying and Using Local Exhaust Ventilation (LEV).* Leaflet Series INDG 408, HSE Books.

HSE (2008b) *Safe Use of Work Equipment. Provision and Use of Work Equipment Regulations. Approved Code of Practice and Guidance,* 3rd edn. Legal Series L22, HSE Books.

HSE (2008c) *Simple Guide to the Provision and Use of Work Equipment Regulations 1998.* Leaflet Series INDG291, HSE Books.

HSE (2008d) *Time to Clear the Air! A Worker's Pocket Guide to Local Exhaust Ventilation (LEV).* Leaflet Series INDG 409, HSE Books.

HSE (2009a) *Working Alone. Health and Safety Guidance on the Risks of Working Alone.* Leaflet Series INDG73 (REV2), HSE Books.

HSE (2009b) *Working with Substances Hazardous to Health. A Brief Guide to COSHH.* Leaflet Series INDG136 (REV3), HSE Books.

HSE (2011a) *Controlling Airborne Contaminants at Work.* A Guide to Local Exhaust Ventilation. Health and Safety Guidance Series HSG258, HSE Books.

HSE (2011b) *Safety in the Installation and Use of Gas Systems and Appliances. The Gas Safety (Installations and Use) Regulations 1998. Approved Code of Practice and Guidance,* 3rd edn. Legal Series L56, HSE Books.

Note: The HSE web site (www.hse.gov.uk) is an invaluable source of information.

Index